Applied
Computational Statistics in
Longitudinal Research

Applied Computational Statistics in Longitudinal Research

Michael J. Rovine
Alexander von Eye

Department of Human Development and Family Studies
College of Health and Human Development
The Pennsylvania State University
University Park, Pennsylvania

ACADEMIC PRESS, INC.
Harcourt Brace Jovanovich, Publishers

Boston   San Diego   New York
London   Sydney   Tokyo   Toronto

Copyright © 1991 by Academic Press, Inc.
All rights reserved.
No part of this publication may be reproduced or
transmitted in any form or by any means, electronic
or mechanical, including photocopy, recording, or
any information storage and retrieval system, without
permission in writing from the publisher.

Cover design by Elizabeth E. Tustian

ACADEMIC PRESS, INC.
1250 Sixth Avenue, San Diego, CA 92101

United Kingdom Edition published by
ACADEMIC PRESS LIMITED
24–28 Oval Road, London NW1 7DX

Library of Congress Cataloging in Publication Data:

Rovine, Michael J.
    Applied computational statistics in longitudinal research /
  Michael J. Rovine, Alexander von Eye.
      p. cm.
    Includes bibliographical references.
    Includes index.
    ISBN 0-12-599450-8 (alk. paper)
    1. Medicine—Longitudinal studies—Statistical methods.
  2. Psychology, Experimental—Longitudinal studies—Statistical
  methods. I. Eye, Alexander von. II. Title.
    [DNLM: 1. Longitudinal Studies. 2. Statistics.   WA 950 R875a]
  R853.S7R68   1990
  610'.72—jc20
  DNLM/DLC
  for Library of Congress   90-844

Printed in the United States of America   CIP
91 92 93 94   9 8 7 6 5 4 3 2 1

To our parents

Contents

Preface

Longitudinal research seems to be constantly acquiring new statistical strategies for the analysis of empirical data. Each new strategy represents a set of skills that, when acquired, can lead either to a more direct test of the hypotheses under consideration or to a better description of the phenomena being explored.

Acquiring those skills can be difficult. Many researchers who are presented with a paper describing the latest technique find it almost impossible to turn that paper into a computer output that describes their data. This often leads to a search for someone who is not only familiar with the technique, but who also has the programming skills (or the requisite knowledge of statistical packages) required to complete the task. This frequently takes the form of one researcher sending the other a sample program which is then "cannibalized" to generate the solution. However, major problems result if the new user has little understanding of the output. In this case, an annotated output can do wonders to help understanding.

Occasionally, one user has access to a program that solves a problem efficiently, but the same software is not available at another's facility. In that case, "cut and paste programming" is not an option. What can often be of use, however, is a complete solution to a known data set along with the data.

If the latter individual is forced to write a program or try out a new package, that procedure can be road-tested on a data set with a known solution.

Such examples are not easily available for procedures involving longitudinal data. This book fills that gap by providing examples of some of the newer statistical strategies in longitudinal research. In this sense, it is intended to be a "how to" book.

The procedures exemplified are those presented in the two volumes of *Statistical Methods in Longitudinal Research*, edited by Alexander von Eye (1990, Academic Press). The techniques discussed can be divided into four general groups:

The first group includes two general techniques in data analysis. These are missing data analysis and repeated measures analysis of variance.

The second group is methods for structuring change. In this group, longitudinal factor analysis, latent variable models of change using structural equation modeling, and scaling techniques will be exemplified.

The third group is time series models. Examples of event history analysis, power series and polynomial growth curves, spectral analysis, Box-Jenkins models, and generalized learning curves (Tucker analysis) are given.

The final group is a selection of techniques for analyzing categorical data. These methods include loglinear modeling, latent class analysis, finite mixture distributions, prediction analysis and configural frequency analysis.

Examples of the above techniques are presented in the following format:

1. A general summary of the method is presented.
2. An outline of the computational procedures is given, including a brief description of the software selected.
3. An annotated data example is presented that describes the data set selected and how to input it. The program for generating the solution is then presented along with an annotated portion of the output.
4. A brief discussion of the results is then presented.

This book has been set up in a way that allows the reasonably knowledgeable user to run some of these relatively fancy procedures rather quickly. Of course, a more thorough understanding of each procedure is necessary for the best interpretation of the results.

This book can also be used as a set of examples to enable the user to understand the techniques described more completely in the two volumes of *Statistical Methods in Longitudinal Research*. However this book is used, we sincerely hope it somewhat eases the pain of getting acquainted with still another statistical technique.

Many people have contributed to the completion of this project. We would like to thank Susan McCulley Gay and Klaus Peters of Academic Press, who guided the process of creating this book to completion. John Nesselroade's Research Methodology Center provided a forum for presenting many of the examples in this book. We would like to thank members of the center's Monday research colloquium for reviewing and commenting on many of the chapters. These include John Nesselroade, Paul Games, Phil Wood, Bruce Pugesek, Verneda Hamm, Adrian Tomer, Jules Thayer, Tom Feagans, Cindy Bergemann, Bob Intrieri, and Jim Reid. Laura Lengnick provided welcomed input regarding problems of repeated measures analysis. Discussions with many of our colleagues in the Department of Human Development and Family Studies and in the College of Health and Human Development helped us frame many of the chapters in a way that we hope will be helpful for users. We thank Phil Wood for providing the data example for the scaling chapter. Rich Lerner generously lent us the use of his PCs and printers. We appreciate the expertise of Donna Ballock, who put our piecemeal manuscript into its final form.

All of the above-mentioned contributed to the quality of this endeavor. Any errors in this book remain the responsibility of the authors.

Michael J. Rovine
Alexander von Eye
University Park, Pennsylvania

Chapter 1

Missing-Data Estimation for Nested Panel Design

1. Summary of Method

When the pattern of missing data is nested in a longitudinal or panel design (that is, participants who drop out of a study stay out of the study at any future wave), maximum-likelihood estimation can be used to generate estimates of the means and the variance/covariance matrix of that data set. A general method for computing the estimation has been presented by Marini, Olsen, and Rubin (1980) and Little and Rubin (1987). The method suggested is easiest to implement when considered as a matrix algebra problem. In this case the computation involves the application of well-known matrix operators to a set of matrices, one calculated at each wave.

The steps required for a satisfactory estimation of the missing data problem require more, however, than just the computation. Missing data presents as much a descriptive problem as it does a computational problem. As a characteristic of the data set, the pattern of missing data affects how the results of any analysis should be interpreted. The following steps are therefore suggested as a way to handle the estimation of missing data in a longitudinal research design:

1. Describe the pattern of missing data. First, generate the frequency distribution for each of the variables of interest for a particular analysis. To as-

1

certain the pattern of missingness in associations, generate a pairwise covariance or correlation matrix along with the Ns for each pair of variables. To determine whether there is anything common about the pattern of missingness across variables, create a dichotomous marker variable for each original variable and correlate these.

2. Estimate a matrix of means and correlations (or covariances) separately on subsets of observations based on everyone who has complete data up to and including every wave. Use characteristics of maximum-likelihood regression to generate estimates based on all of the available data.

3. Compare the estimated matrix with boundary values determined using pairwise and listwise deletion for all variables of interest across all occasions.

4. If necessary, estimate individual data points based on the maximum-likelihood matrix.

5. Complete the analysis using the new matrix. Compare results against the same analysis using complete data and look for any discrepancies.

2.   Outline of the Computational Procedures

This section outlines steps for completing the estimation. For ease of computation, we have selected SAS as a preprocessor and BMDPAM (BMDP, 1985) for estimation of the descriptive characteristics of the missing data pattern. As of this writing, SAS includes no generalized missing data estimation procedure. Many installations allow the user to call BMDP procedures as SAS PROCS (PROC BMDP). We take advantage of this option in the following example.

BMDPAM is a general purpose missing data estimation program. It generates a number of diagnostics to help discern the pattern of missing data. A number of estimation options are available. Of particular interest is the EM option (see Rovine and Delaney, 1990). This option uses an interative maximum-likelihood algorithm to estimate a vector of means and a matrix of association. It can also estimate individual data points and include those estimates in an output file. We use the program both to establish descriptive characteristics of the data set and to evaluate the pattern of missingness.

The program cannot readily handle longitudinal data. As a result, we carry out the longitudinal maximum-likelihood estimation using SAS Interactive Matrix Language (IML). IML is the matrix language module of SAS (SAS PROC IML, 1985). It is a multi-purpose program that can be used to write procedures for any problem that be expressed in matrix terms. IML can either be used interactively, or the statements can be stored and submitted as a single EXEC.

The input to the IML program consists of a set of covariance matrices augmented by a vector of means, one for each wave including those who have complete data up to and including that wave. The procedure uses a series of maximum-likelihood regressions to come up with an estimated matrix of covariances and means. The mechanics of the procedure have been detailed elsewhere (Marini *et al.* 1980; Rovine and Delaney, 1990).

The analyses reported in this chapter were conducted on the mainframe. The same procedures could be adapted to the microcomputer, provided the software has the capacity to operate on matrices.

3.   An Annotated Data Example

The data selected for this example come from three waves of the Child–Family Development Project (Belsky *et al.* 1984). Variables to be included in the analysis are four subscales from the Braiker–Kelley marital satisfaction scale: LOVE, CONFlict, MAINTenance, and AMBivalence.

Step 1:   Data Input

Using SAS as a preprocessor, we read in a set of raw data as follows:

```
OPTIONS NOCENTER LINESIZE=80 NONUMBER NODATE;
DATA MISEXAMP;
INFILE MISEXAMP;
INPUT ID      PWLOVE      PWCONF      PWMAINT      PWAMB
      W3LOVE      W3CONF      W3MAINT      W3AMB
      W9LOVE      W9CONF      W9MAINT      W9AMB;
```

As can be seen, repeatedly measured variables carry the same stem describing the variable with a prefix indicating the wave of data. This input format considers a case to be a complete set of data across all waves with the individual as the unit of analysis.

Step 2:   Determination of Descriptive Characteristics of Data

SAS is first used to generate descriptive characteristics of each variable. Missing data for each variable can be determined. In particular, scales created by compositing items are checked to see whether the complete scale is missing or whether one or just a few of the items are missing. In the latter case, existent information can be used to generate a scale score (Rovine and Delaney, 1990). A sample set of commands for "inflating" existent items is

included below. Once we have cleaned the data in this fashion, we can turn
to the generation of the description of the pattern of missing data. One way
to check whether the pattern is nested is to create a pattern variable for each
observation as follows:

```
ARRAY VARS PWLOVE PWCONF PWMAINT PWAMB W3LOVE W3CONF W3MAINT
    W3AMB W9LOVE W9CONF W9MAINT W9AMB;
ARRAY VARMIS DLOVE1 DCONF1 DMAINT1 DAMB1 DLOVE2 DCONF2 DMAINT2
    DAMB2 DLOVE3 DCONF3 DMAINT3 DAMB3;
DO OVER VARMIS;
    VARMIS=1;
    END;
DO OVER VARMIS;
    IF VARS EQ . THEN VARMIS=0;
    END;
VARPATRN=100*DLOVE1+10*DLOVE2+DLOVE3;
```

Any pattern value in which a "0" is followed by a "1" represents a violation
of the nested pattern. If such violations are few in number, a reasonable
strategy would eliminate those cases from the particular analysis. Little and
Rubin (1987), though, present a modification of the method to be described
that can handle the non-nested pattern.

BMDP programs can be called as PROCs in a SAS run using the PROC
BMDP; command. We use this facility to execute the following program:

```
PROC FREQ;
    TABLES VARPATRN;

PROC BMDP PROG=BMDPAM;
    PARMCARDS;
/INPUT UNIT=03. CODE='MISEXAMP'.
/VARIABLE USE=PWLOVE,PWCONF,PWMAINT,PWAMB,
                W3LOVE,W3CONF,W3MAINT,W3AMB,
                W9LOVE,W9CONF,W9MAINT,W9AMB.
/ESTIMATE TYPE=ML.
    METHODS=REGR.
/PRINT
/END.
```

This run provides us with most of the descriptive information necessary to
prepare and interpret the estimation runs. The output from the run follows:

VARPATRN	FREQUENCY	PERCENT	CUMULATIVE FREQUENCY	CUMULATIVE PERCENT
100	10	5.9	10	5.9
101	3	1.8	13	7.6
110	11	6.5	24	14.1
111	146	85.9	170	100.0

The first part of the output represents a SAS PROC FREQ run on the pattern variable. As can be seen, three cases violate the nested pattern. For the
purpose of the maximum-likelihood estimates, these three cases can be de-
leted. The remaining three "legal" patterns can be used to generate the correlation matrices that will be used for the estimation.

The BMDPAM program first indicates the options selected.

```
GROUPING VARIABLE . . . . . . . . . . . . . . .
PRECISION . . . . . . . . . . . . . . . . . . . DOUBLE
TOLERANCE FOR MATRIX INVERSION. . . . . . . . 0.0001000
ESTIMATION METHOD . . . . . . . . . . . . . . . REG
F TO ENTER LIMIT. . . . . . . . . . . . . . . .     4.00
MISSING VARIABLE LIMIT PER CASE . . . . . . . .        6
MISSING CASE LIMIT PER VARIABLE . . . . . . . .   50.00 PERCENT
EIGENVALUE LOWER LIMIT. . . . . . . . . . .        0.00100
RIDGE PARAMETER . . . . . . . . . . . . . . .     1.00000
METHOD OF COMPUTING COVARIANCE. . . . . . . . .MAX.LIK.
MAXIMUM NUMBER OF ITERATIONS FOR MAX. LIKE. . .      10
CONVERGENCE CRITERION FOR MAXIMUM LIKELIHOOD. . 0.01000
```

Most of the selections shown are by default. The F-to enter refers to an option which selects as predictors of the missing variable all other variables that meet the criterion represented by the *F*-value. The program allows the user to set the maximum number of missing cases allowed for any variable. The number of missing variables required to delete a case can also be set. If the estimation methods generate a matrix in which pairs of the variables have high correlations due to the estimation procedure, multicollinearity may be a problem. To indicate the degree of dependence among variables, the program checks for small or negative eigenvalues (an indication of dependence among variables). The user can decide what constitutes a small eigenvalue, or as we did, let the computer select the value.

If there is a problem with multicollinearity, one solution that has been suggested is the use of ridge regression (Affifi and Clark, 1984). The ridge parameter can be selected. Often values above .2 add little to the solution.

Next, the BMDP output presents a useful summary of missing data characteristics. Frequencies and percentages of missing data are presented for each variable.

```
NUMBER OF CASES READ. . . . . . . . . . . . . . . . .    170
** CASES WITH TOO MANY VARIABLES MISSING OR BEYOND LIMITS   10
       REMAINING NUMBER OF CASES . . . . . . . . . . .    160

   NUMBER OF CASES WITH NO DATA MISSING AND WITH
       POSITIVE CASE WEIGHT. . . . . . . . . . . . . .    146

** THESE CASES ARE NOT USED IN ESTIMATING MISSING DATA.

PERCENTAGES OF MISSING DATA FOR EACH VARIABLE IN EACH GROUP
-----------------------------------------------------------

THESE PERCENTAGES ARE BASED ON SAMPLE SIZES AND GROUP SIZES
REPORTED WITH THE UNIVARIATE SUMMARY STATISTICS BELOW.
VARIABLES WITHOUT MISSING DATA ARE NOT INCLUDED.

W3LOVE      6      1.9
W3CONF      7      1.9
W3MAINT     8      1.9
W3AMB       9      1.9
W9LOVE     10      6.9
W9CONF     11      6.9
W9MAINT    12      6.9
W9AMB      13      6.9
```

These are followed by a set of descriptive statistics.

UNIVARIATE SUMMARY STATISTICS

VARIABLE	SAMPLE SIZE	MEAN	STANDARD DEVIATION	COEFFICIENT OF VARIATION	SMALLEST VALUE
2 PWLOVE	160	80.98750	7.19712	0.088867	50.00000
3 PWCONF	160	20.09375	5.93974	0.295601	6.00000
4 PWMAINT	160	32.18750	5.93485	0.184384	14.00000
5 PWAMB	160	10.04375	4.70026	0.467978	5.00000
6 W3LOVE	157	78.01911	9.06774	0.116225	28.00000
7 W3CONF	157	20.85987	6.42357	0.307939	5.00000
8 W3MAINT	157	29.99363	6.18051	0.206061	9.00000
9 W3AMB	157	10.52229	5.48843	0.521600	5.00000
10 W9LOVE	149	76.43624	9.59048	0.125470	26.00000
11 W9CONF	149	21.71812	6.00684	0.276582	8.00000
12 W9MAINT	149	28.59060	6.42421	0.224696	13.00000
13 W9AMB	149	11.66443	5.94867	0.509983	5.00000

A matrix is then presented which shows the common sample size for each pair of variables. This can be used as an indication of how severe missing data is across waves.

SAMPLE SIZES FOR EACH PAIR OF VARIABLES

(NUMBER OF TIMES BOTH VARIABLES ARE AVAILABLE)
IN ORDER TO SAVE SPACE, VARIABLES WITH NO MISSING
DATA OR THAT HAVE NO DATA ARE NOT INCLUDED.

		W3LOVE	W3CONF	W3MAINT	W3AMB	W9LOVE	W9CONF	W9MAINT	W9AMB
		6	7	8	9	10	11	12	13
W3LOVE	6	157							
W3CONF	7	157	157						
W3MAINT	8	157	157	157					
W3AMB	9	157	157	157	157				
W9LOVE	10	146	146	146	146	149			
W9CONF	11	146	146	146	146	149	149		
W9MAINT	12	146	146	146	146	149	149	149	
W9AMB	13	146	146	146	146	149	149	149	149

If only a couple of cases are missing, it may be appropriate to use listwise deletion to create a vector of means and a variance/covariance matrix. The missing data matrix is then re-expressed in terms of percentages of missing data for each pair of variables.

PAIRWISE PERCENTAGES OF MISSING DATA

DIAGONAL ELEMENTS ARE THE PERCENTAGES THAT EACH VARIABLE
IS MISSING. OFF-DIAGONAL ELEMENTS ARE THE PERCENTAGES
EITHER VARIABLE IS MISSING. THESE PERCENTAGES DO NOT INCLUDE
CASES WITH MISSING GROUP OR WEIGHT VARIABLES, CASES WITH
ZERO WEIGHTS, CASES EXCLUDED BY SETTING USE EQUAL TO A
NON-POSITIVE VALUE BY TRANSFORMATIONS, OR CASES WITH GROUPING
VALUES NOT USED. VARIABLES WITH NO MISSING DATA OR THAT
HAVE NO DATA ARE NOT INCLUDED HERE.

		W3LOVE	W3CONF	W3MAINT	W3AMB	W9LOVE	W9CONF	W9MAINT	W9AMB
		6	7	8	9	10	11	12	13
W3LOVE	6	1.9							
W3CONF	7	1.9	1.9						
W3MAINT	8	1.9	1.9	1.9					
W3AMB	9	1.9	1.9	1.9	1.9				
W9LOVE	10	8.7	8.7	8.7	8.7	6.9			
W9CONF	11	8.7	8.7	8.7	8.7	6.9	6.9		
W9MAINT	12	8.7	8.7	8.7	8.7	6.9	6.9	6.9	
W9AMB	13	8.7	8.7	8.7	8.7	6.9	6.9	6.9	6.9

A correlation matrix of dichotomized variables (0=missing, 1=present) comes next.

```
CORRELATIONS OF THE DICHOTOMIZED VARIABLES
-------------------------------------------
WHERE FOR EACH VARIABLE ZERO INDICATES THAT THE VALUE WAS
MISSING AND ONE INDICATES THAT THE VALUE WAS PRESENT.
VARIABLES WITH NO MISSING DATA OR THAT ARE COMPLETELY
MISSING ARE NOT INCLUDED.
```

		W3LOVE	W3CONF	W3MAINT	W3AMB	W9LOVE	W9CONF	W9MAINT	W9AMB
		6	7	8	9	10	11	12	13
W3LOVE	6	1.000							
W3CONF	7	1.000	1.000						
W3MAINT	8	1.000	1.000	1.000					
W3AMB	9	1.000	1.000	1.000	1.000				
W9LOVE	10	-0.038	-0.038	-0.038	-0.038	1.000			
W9CONF	11	-0.038	-0.038	-0.038	-0.038	1.000	1.000		
W9MAINT	12	-0.038	-0.038	-0.038	-0.038	1.000	1.000	1.000	
W9AMB	13	-0.038	-0.038	-0.038	-0.038	1.000	1.000	1.000	1.000

Correlations of 1.00 between any pair of dichotomous variables mean that the missing data pattern is the same, so if the only cause of missing data is dropouts from the study, the portion of the matrix representing a set of variables within a wave should be all 1.00s. Correlation between waves indicates the degree to which the same underlying process is generating the missing data. Low between-wave correlations (here $r = -.038$) indicate independent missing data patterns probably representing an underlying random process. Following this matrix is a listing of which variables are missing for each case.

```
PATTERN OF MISSING DATA AND DATA BEYOND LIMITS
----------------------------------------------
COUNT OF MISSING VARIABLES INCLUDES DATA BEYOND LIMITS.
THE COLUMN LABELED WT. IS FOR THE CASE WEIGHT, IF ANY.
M REPRESENTS A MISSING VALUE.  B REPRESENTS A VALUE GREATER
THAN THE MAXIMUM LIMIT.  S REPRESENTS A VALUE LESS THAN THE
MINIMUM LIMIT.
```

```
                                    P P P P W W W W W W W
                                    W W W W 3 3 3 3 9 9 9
                                    L C M A L C M A L C M A
                                    O O A M O O A M O O A M
                                    V N I B V N I B V N I B
                          NO OF      W E F N   E F N   E F N
        CASE    CASE MISS.           T     T     T     T
        LABEL   NO.  VARS.  GROUP

                 22      8           M M M M M M M M
                 30      8           M M M M M M M M
                 49      4                   M M M M
                 50      4                   M M M M
                 51      4                   M M M M
                 53      4                   M M M M
                 54      4                   M M M M
                 59      4                   M M M M
                 73      8           M M M M M M M M
                 77      8           M M M M M M M M
                 78      4           M M M M
                 83      8           M M M M M M M M
                 88      8           M M M M M M M M
                 90      4                   M M M M
                105      8           M M M M M M M M
                111      4                   M M M M
                113      4                   M M M M
```

```
117   4                        M M M M
120   8          M M M M M M M M
157   8          M M M M M M M M
158   8          M M M M M M M M
165   4                        M M M M
166   4            M M M M
168   4            M M M M
```

Maximum-likelihood estimation of the vector of means and the correlation matrix are computed using an EM algorithm. In the case of cross-sectional data, these represent good estimates which tend to fall somewhere between a matrix generated by listwise deletion and one generated by pairwise deletion. In the case of longitudinal data, however, these estimates may be problematic. Using regression estimates, these variables allow later waves to contribute the estimation of missing data for prior waves. As a result, in the case of longitudinal data, we recommend bypassing these estimates and moving on to a method that makes use of time-ordered characteristics of the data. These estimates are included for purposes of comparison with the panel estimates to be discussed.

```
MAXIMUM LIKELIHOOD ESTIMATES
----------------------------
```

VARIABLE	MEAN	STANDARD DEVIATION
2 PWLOVE	80.98750	7.19712
3 PWCONF	20.09375	5.93974
4 PWMAINT	32.18750	5.93485
5 PWAMB	10.04375	4.70026
6 W3LOVE	78.14883	9.08237
7 W3CONF	20.81686	6.40625
8 W3MAINT	30.06994	6.19139
9 W3AMB	10.50101	5.48948
10 W9LOVE	76.14148	9.69904
11 W9CONF	21.91118	6.13766
12 W9MAINT	28.59950	6.39742
13 W9AMB	11.84707	6.09072

```
CORRELATIONS
------------
```

		PWLOVE	PWCONF	PWMAINT	PWAMB	W3LOVE	W3CONF	W3MAINT	W3AMB
		2	3	4	5	6	7	8	9
PWLOVE	2	1.000							
PWCONF	3	-0.391	1.000						
PWMAINT	4	0.473	-0.036	1.000					
PWAMB	5	-0.622	0.506	-0.144	1.000				
W3LOVE	6	0.695	-0.375	0.320	-0.524	1.000			
W3CONF	7	-0.312	0.664	0.040	0.429	-0.430	1.000		
W3MAINT	8	0.390	0.012	0.676	-0.117	0.480	0.075	1.000	
W3AMB	9	-0.489	0.427	-0.121	0.674	-0.714	0.520	-0.204	1.000
W9LOVE	10	0.739	-0.324	0.311	-0.577	0.788	-0.314	0.379	-0.598
W9CONF	11	-0.355	0.648	-0.038	0.447	-0.353	0.604	0.060	0.438
W9MAINT	12	0.360	0.084	0.644	-0.140	0.316	0.076	0.715	-0.122
W9AMB	13	-0.358	0.406	-0.083	0.592	-0.408	0.348	-0.044	0.654

```
                W9LOVE   W9CONF   W9MAINT   W9AMB

                  10       11       12      13
W9LOVE    10    1.000
W9CONF    11   -0.395    1.000
W9MAINT   12    0.411    0.081    1.000
W9AMB     13   -0.609    0.545   -0.133    1.000
```

```
EIGENVALUES OF CORRELATION MATRIX
----------------------------------

    5.2146     0.7461     0.3711     0.1672
    2.6183     0.5774     0.3203     0.1448
    0.9651     0.5216     0.2645     0.0889
```

```
SQUARED MULTIPLE CORRELATIONS OF EACH VARIABLE WITH ALL OTHER VARIABLES
----------------------------------------------------------------------
(MEASURES OF MULTICOLLINEARITY OF VARIABLES)
AND TESTS OF SIGNIFICANCE OF MULTIPLE REGRESSION
DEGREES OF FREEDOM FOR F-STATISTICS ARE      11 AND    148
```

VARIABLE NO.	NAME	R-SQUARED	F-STATISTIC	SIGNIFICANCE (P LESS THAN)
2	PWLOVE	0.735009	37.32	0.00000
3	PWCONF	0.590592	19.41	0.00000
4	PWMAINT	0.598580	20.06	0.00000
5	PWAMB	0.643581	24.29	0.00000
6	W3LOVE	0.825890	63.82	0.00000
7	W3CONF	0.594150	19.70	0.00000
8	W3MAINT	0.696459	30.87	0.00000
9	W3AMB	0.756872	41.88	0.00000
10	W9LOVE	0.818846	60.82	0.00000
11	W9CONF	0.577811	16.41	0.00000
12	W9MAINT	0.639401	23.86	0.00000
13	W9AMB	0.696824	30.92	0.00000

BMDPAM also provides the eigenvalues of the estimated matrix, along with multicollinearity tests for each of the variables. As mentioned above, these help to indicate the degree to which the estimation of missing data has increased dependencies among variables. If a subset of variables has the same pattern of missingness and a large number of estimated values, correlations between pairs of those variables become inflated. The squared multiple correlations computed can help determine how well each variable can be predicted by all others in the data set. If the SMC values are large, the matrix estimated could be improper (van Driel, 1978).

BMDPAM generates estimates of individual missing-data points based on the maximum-likelihood estimates just reported. These values can be used to "fill in" missing data in the original data set. A few of these lines follow.

```
IN THE TABLE BELOW, ESTIMATES THAT ARE LESS THAN THE MINIMUMS
STATED IN THE VARIABLE PARAGRAPH ARE FLAGGED BY THE
LETTER 'S' (SMALL) AFTER THE ESTIMATE.  ESTIMATES
GREATER THAN THE MAXIMUMS ARE FLAGGED BY THE LETTER
'B' (BIG).
MAHALANOBIS DISTANCES ARE COMPUTED FROM EACH CASE TO THE
```

CENTROID OF ITS GROUP. ONLY THOSE VARIABLES WHICH WERE
ORIGINALLY AVAILABLE ARE USED--ESTIMATED VALUES ARE NOT USED.
FOR LARGE MULTIVARIATE NORMAL SAMPLES, THE MAHALANOBIS
DISTANCES HAVE AN APPROXIMATELY CHI-SQUARE DISTRIBUTION WITH
THE NUMBER OF DEGREES OF FREEDOM EQUAL TO THE NUMBER OF
NONMISSING VARIABLES. SIGNIFICANCE LEVELS REPORTED BELOW
THAT ARE LESS THAN .001 ARE FLAGGED WITH AN ASTERISK.

CASE NUMBER	MISSING VARIABLE	ESTIMATE	R-SQUARED	GROUP	CHI-SQ	CHISQ
49	W9LOVE	81.2473	0.706		1.914	8
49	W9CONF	20.8799	0.499		1.914	8
49	W9MAINT	33.6329	0.581		1.914	8
49	W9AMB	9.6815	0.498		1.914	8
50	W9LOVE	67.2262	0.706		28.302	8
50	W9CONF	28.8088	0.499		28.302	8
50	W9MAINT	27.8377	0.581		28.302	8
50	W9AMB	14.1539	0.498		28.302	8
51	W9LOVE	72.1296	0.706		2.064	8
51	W9CONF	23.3477	0.499		2.064	8
51	W9MAINT	28.5222	0.581		2.064	8
51	W9AMB	12.5776	0.498		2.064	8
52					11.849	12
53	W9LOVE	73.1827	0.706		1.156	8
53	W9CONF	23.0371	0.499		1.156	8
53	W9MAINT	26.8636	0.581		1.156	8
53	W9AMB	13.3693	0.498		1.156	8

If the estimated value for a particular variable is either very large or very small compared to the rest of that variable's distribution, BMDPAM flags the value.

Step 3:   Preparation of Matrices for the Maximum-Likelihood Estimation

The data is first sorted by pattern of missing data. For the three waves of data the nested pattern only allow the following legal patterns (subgroup size in parentheses): $111(n_1=146)$, $110(n_2=11)$, and $100(n_3=10)$ representing those with complete data at all three occasions, those who dropped out after two occasions, and those who dropped out after the first occasion. The first wave data is used to calculate a correlation matrix, **MATRIX1**, and a vector of means, **MEANS1**, on all cases with complete data for first-wave variables $(n_1+n_2+n_3=167)$ which, because of the nested requirement, represents all in the study with legal patterns. A second matrix and vector are then calculated on the wave 1 and wave 2 variables for all those with complete data at waves 1 and 2 $(n_1+n_2=157)$. Finally, a third matrix and vector are calculated on wave 1, wave 2, and wave 3 variables for all having complete data on all three waves $(n_1=146)$. These matrices and vectors were calculated by first sorting according to the pattern variable and then using SAS PROC CORR to compute the matrices. The resultant matrices appear along with program input below.

Our data example is based on a correlation matrix. An alternative analysis would make use of a covariance matrix (COV option in PROC CORR). Although covariance matrices are generally more appropriate for looking at longitudinal data (especially when the change in variance of measures over time is of interest), we selected the correlation metric to make it somewhat easier to compare the differences between the original (complete data) and estimated coefficients. The procedures discussed below would, however, hold for a covariance matrix.

Step 4: Estimation of the Matrix of Means and Correlations

The program used to estimate the matrix of means and correlations is written in the SAS matrix language IML. Although input vectors and matrices could be called directly from SAS DATA and PROC procedures, we include the input as part of the program for descriptive purposes. The estimation procedure is basically a set of sweeps and reverse sweeps on separate matrices representing complete data up to and including each wave (Marini *et al.* 1980; Dempster, 1969). For the three waves of data analyzed below, this requires three input matrices. Since the variables we are dealing with have been standardized, means will be 0, and the matrix of associations will be a correlation matrix. The same procedure, however, would work with non-zero means and a covariance matrix as input. The program follows:

```
PROC IML;
START;
CONSTANT=1;

***                                          ;
*** WAVE 1 MEANS AND VARIANCE/COVARIANCE MATRIX;
***                                          ;

MEANS1={0.0 0.0 0.0 0.0};

COV1={1.000 -.386 .445 -.626,
      -.386 1.000 .021 .509,
       .445 .021 1.000 -.127,
      -.626 .509 -.127 1.000};

***                              ;
*** CONCATENATE WAVE 1 MATRIX    ;
***                              ;

A1=CONSTANT||MEANS1;
A2=MEANS1`||COV1;
A=A1//A2;
PRINT A;
ACAT=A;
```

This part of the program takes a vector of means for wave 1 (**MEANS 1**) and a correlation matrix (**COV1**) and shapes a matrix **A** with a constant=1 in the

(1,1) position and the vector of means occupying both the first column and the first row. The next part of the program first sweeps the matrix (Dempster, 1969) on the constant and then all on all wave 1 variables.

```
***                  ;
*** SWEEP WAVE 1 MATRIX;
***                  ;

ASWEPT={0 0 0 0 0,
        0 0 0 0 0,
        0 0 0 0 0,
        0 0 0 0 0,
        0 0 0 0 0};
ASWEEPST=1;
ASWEEPEN=5;
ROWS=5;
COLUMNS=5;
DO K=ASWEEPST TO ASWEEPEN;
 ASWEPT(!K,K!)=-1/ACAT(!K,K!);
DO I=1 TO ROWS;
 IF I=K THEN GOTO CONTIN2;
 ASWEPT(!I,K!)=ACAT(!I,K!)/ACAT(!K,K!);
 CONTIN2: COMMENT;
DO J=1 TO COLUMNS;
 IF J=K THEN GOTO CONTIN3;
 ASWEPT(!K,J!)=ACAT(!K,J!)/ACAT(!K,K!);
 CONTIN3: COMMENT;
 IF I=K THEN GOTO CONTIN1;
 IF J=K THEN GOTO CONTIN1;
 ASWEPT(!I,J!)=ACAT(!I,J!)-ACAT(!I,K!)#ACAT(!K,J!)/ACAT(!K,K!);
 CONTIN1: COMMENT;
END;
END;
ACAT=ASWEPT;
END;
PRINT ASWEPT;
ACAT=ASWEPT;
PRINT ACAT;
```

The results of the sweep, **ASWEPT** are then passed along to the next step in which means and correlations of complete data through wave 2 (**MEANS2, COV2**) are shaped into matrix **B**. **B** is then swept on first the wave 1 variables and then the wave 2 variables which generate the matrix **BSWEPT**. **BSWEPT** is then combined with **ASWEPT** to form the matrix **BCAT**. This matrix is then passed on to the next step.

```
***                  ;
*** WAVE 2 MEANS AND VARIANCE/COVARIANCE MATRIX;
***                  ;

MEANS2={0.0 0.0 0.0 0.0 0.0 0.0 0.0 0.0};

COV2={1.000 -.382 .465 -.627 .692 -.308 .382 -.488,
      -.382 1.000 -.023 .518 -.368 .666 .023 .430,
      .465 -.023 1.000 -.141 .311 .047 .675 -.115,
      -.627 .518 -.141 1.000 -.527 .432 -.105 .673,
      .692 -.368 .311 -.527 1.000 -.427 .472 -.714,
      -.308 .666 .047 .432 -.427 1.000 .083 .519,
      .382 .023 .675 -.105 .472 .083 1.000 -.195,
      -.488 .430 -.115 .673 -.714 .519 -.195 1.000};
```

```
***                          ;
*** CONCATENATE WAVE 2 MATRIX;
***                          ;

B1=CONSTANT||MEANS2;
B2=MEANS2`||COV2;
B=B1//B2;
PRINT B;
BCAT=B;

***                    ;
*** SWEEP WAVE 2 MATRIX;
***                    ;

BSWEPT={0 0 0 0 0 0 0 0 0,
        0 0 0 0 0 0 0 0 0,
        0 0 0 0 0 0 0 0 0,
        0 0 0 0 0 0 G 0 0,
        0 0 0 0 0 0 0 0 G,
        0 0 0 0 0 0 0 0 0,
        0 0 0 0 0 0 0 0 0,
        0 0 0 0 0 0 0 0 0,
        0 0 0 0 0 0 0 0 0};
BSWEEPST=1;
BSWEEPEN=9;
ROWS=9;
COLUMNS=9;
DO K=BSWEEPST  TO BSWEEPEN;
 BSWEPT(|K,K|)=-1/BCAT(|K,K|);
DO I=1 TO ROWS;
 IF I=K THEN GOTO CONTIN4;
 BSWEPT(|I,K|)=BCAT(|I,K|)/BCAT(|K,K|);
 CONTIN4: COMMENT;
DO J=1 TO COLUMNS;
 IF J=K THEN GOTO CONTIN5;
 BSWEPT(|K,J|)=BCAT(|K,J|)/BCAT(|K,K|);
 CONTIN5: COMMENT;
 IF I=K THEN GOTO CONTIN6;
 IF J=K THEN GOTO CONTIN6;
 BSWEPT(|I,J|)=BCAT(|I,J|)-BCAT(|I,K|)#BCAT(|K,J|)/BCAT(|K,K|);
 CONTIN6: COMMENT;
END;
END;
BCAT=BSWEPT;
END;
PRINT BSWEPT "RUN";
BCAT=BSWEPT;
PRINT BCAT;
BCAT21=BSWEPT(|{6 7 8 9},{1 2 3 4 5}|);
BCAT12=BCAT21`;
BCAT22=BSWEPT(|{6 7 8 9},{6 7 8 9}|);
BCAT1=ASWEPT||BCAT12;
BCAT2=BCAT21||BCAT22;
BCAT=BCAT1//BCAT2;
PRINT BCAT21;
PRINT BCAT12;
PRINT BCAT22;
PRINT BCAT;
```

Finally, a matrix **C** is created by shaping the complete data vector of means, **MEANS1**, and correlations, **COV3**, for all three waves of data. This matrix is first swept on first the constant, then the wave 1, and finally the wave 2 variables to create the matrix **CSWEPT**. This is combines with **BSWEPT** to form the matrix **CCAT** which in turn is reverse-swept on the wave 1 and wave 2 variables. This final step generates the maximum-likelihood estimates.

```
***                                           ;
*** WAVE 3 MEANS AND VARIANCE/COVARIANCE MATRIX;
***                                           ;

MEANS3={0.0 0.0 0.0 0.0 0.0 0.0 0.0 0.0 0.0 0.0 0.0 0.0 0.0};

COV3={1.00 -.370 .442 -.656 .702 -.287 .405 -.493 .738 -.336 .362 -.358,
    -.370 1.000 .001 .471 -.316 .651 .038 .355 -.276 .630 .100 .356,
    .442 .001 1.000 -.161 .303 .066 .684 -.120 .289 -.016 .653 -.084,
    -.656 .471 -.161 1.000 -.513 .399 -.123 .642 -.575 .410 -.141 .563,
    .702 -.316 .303 -.513 1.000 -.412 .473 -.701 .782 -.324 .314 -.385,
    -.287 .651 .066 .399 -.412 1.000 .084 .491 -.287 .592 .087 .315,
    .405 .038 .684 -.123 .473 .084 1.000 -.211 .376 .071 .719 -.046,
    -.493 .355 -.120 .642 -.701 .491 -.211 1.000 -.586 .394 -.128 .631,
    .738 -.276 .289 -.575 .782 -.287 .376 -.586 1.000 -.367 .408 -.601,
    -.336 .630 -.016 .410 -.324 .592 .071 .394 -.367 1.000 .093 .515,
    .362 .100 .653 -.141 .314 .087 .719 -.128 .408 .093 1.000 -.140,
    -.358 .356 -.084 .563 -.385 .315 -.046 .631 -.601 .515 -.140 1.000};

***                        ;
*** CONCATENATE WAVE 3 MATRIX;
***                        ;

C1=CONSTANT||MEANS3;
C2=MEANS3`||COV3;
C=C1//C2;
CCAT=C;
PRINT C;

***                  ;
*** SWEEP WAVE 3 MATRIX;
***                  ;

CSWEPT={0 0 0 0 0 0 0 0 0 0 0 0 0,
        0 0 0 0 0 0 0 0 0 0 0 0 0,
        0 0 0 0 0 0 0 0 0 0 0 0 0,
        0 0 0 0 0 0 0 0 0 0 0 0 0,
        0 0 0 0 0 0 0 0 0 0 0 0 0,
        0 0 0 0 0 0 0 0 0 0 0 0 0,
        0 0 0 0 0 0 0 0 0 0 0 0 0,
        0 0 0 0 0 0 0 0 0 0 0 0 0,
        0 0 0 0 0 0 0 0 0 0 0 0 0,
        0 0 0 0 0 0 0 0 0 0 0 0 0,
        0 0 0 0 0 0 0 0 0 0 0 0 0,
        0 0 0 0 0 0 0 0 0 0 0 0 0,
        0 0 0 0 0 0 0 0 0 0 0 0 0};
CSWEEPST=1;
CSWEEPEN=9;
ROWS=13;
COLUMNS=13;
DO K=CSWEEPST TO CSWEEPEN;
 CSWEPT(|K,K|)=-1/CCAT(|K,K|);
DO I=1 TO ROWS;
 IF I=K THEN GOTO CONTIN7;
 CSWEPT(|I,K|)=CCAT(|I,K|)/CCAT(|K,K|);
 CONTIN7: COMMENT;
DO J=1 TO COLUMNS;
 IF J=K THEN GOTO CONTIN8;
 CSWEPT(|K,J|)=CCAT(|K,J|)/CCAT(|K,K|);
 CONTIN8: COMMENT;
 IF I=K THEN GOTO CONTIN9;
 IF J=K THEN GOTO CONTIN9;
 CSWEPT(|I,J|)=CCAT(|I,J|)-CCAT(|I,K|)#CCAT(|K,J|)/CCAT(|K,K|);
 CONTIN9: COMMENT;
END;
END;
CCAT=CSWEPT;
END;
PRINT CSWEPT;
CCAT=CSWEPT;
```

```
CCAT21=CSWEPT(![10 11 12 13],[1 2 3 4 5 6 7 8 9]!);
CCAT12=CCAT21`;
CCAT22=CSWEPT(![10 11 12 13],[10 11 12 13]!);
CCAT11=BSWEPT;
PRINT CCAT11;
PRINT CCAT12;
PRINT CCAT21;
PRINT CCAT22;
CCAT1=CCAT11!!CCAT12;
CCAT2=CCAT21!!CCAT22;
CCAT=CCAT1//CCAT2;
PRINT CCAT;

***                              ;
*** REVERSE SWEEP FINAL MATRIX;
***                              ;

REVSWP={0 0 0 0 0 0 0 0 0 0 0 0 0,
        0 0 0 0 0 0 0 0 0 0 0 0 0,
        0 0 0 0 0 0 0 0 0 0 0 0 0,
        0 0 0 0 0 0 0 0 0 0 0 0 0,
        0 0 0 0 0 0 0 0 0 0 0 0 0,
        0 0 0 0 0 0 0 0 0 0 0 0 0,
        0 0 0 0 0 0 0 0 0 0 0 0 0,
        0 0 0 0 0 0 0 0 0 0 0 0 0,
        0 0 0 0 0 0 0 0 0 0 0 0 0,
        0 0 0 0 0 0 0 0 0 0 0 0 0,
        0 0 0 0 0 0 0 0 0 0 0 0 0,
        0 0 0 0 0 0 0 0 0 0 0 0 0,
        0 0 0 0 0 0 0 0 0 0 0 0 0};
REVSWPST=2;
REVSWPEN=9;
ROWS=13;
COLUMNS=13;
DO K=REVSWPST TO REVSWPEN;
REVSWP(!K,K!)=1/CCAT(!K,K!);
REVSWP(!K,K!)=-REVSWP(!K,K!);
DO I=1 TO ROWS;
IF I=K THEN GOTO CONTIN10;
REVSWP(!I,K!)=-CCAT(!I,K!)/CCAT(!K,K!);
CONTIN10: COMMENT;
DO J=1 TO COLUMNS;
IF J=K THEN GOTO CONTIN11;
REVSWP(!K,J!)=-CCAT(!K,J!)/CCAT(!K,K!);
CONTIN11. COMMENT;
IF I=K THEN GOTO CONTIN12;
IF J=K THEN GOTO CONTIN12;
REVSWP(!I,J!)=CCAT(!I,J!)-CCAT(!I,K!)#CCAT(!K,J!)/CCAT(!K,K!);
CONTIN12: COMMENT;
END;
END;
CCAT=REVSWP;
END;
PRINT REVSWP;

***                                    ;
*** PRINT MAXIMUM-LIKELIHOOD ESTIMATES;
***                                    ;

MAXEST=REVSWP;
PRINT MAXEST;
FINISH;
RUN;
```

The output of this program is a set of matrices representing transitional steps in the procedure. These can act as a check of whether the final estimates are reasonable. A portion of this output follows:

A	COL1	COL2	COL3	COL4	COL5
ROW1	1.0000	0	0	0	0
ROW2	0	1.0000	-0.3860	0.4450	-0.6260
ROW3	0	-0.3860	1.0000	0.0210	0.5090
ROW4	0	0.4450	0.0210	1.0000	-0.1270
ROW5	0	-0.6260	0.5090	-0.1270	1.0000

ACAT	COL1	COL2	COL3	COL4	COL5
ROW1	-1.0000	0	0	0	0
ROW2	0	-2.1790	-0.2946	0.8351	-1.1080
ROW3	0	-0.2946	-1.4032	0.2316	0.5592
ROW4	0	0.8351	0.2316	-1.3468	0.2338
ROW5	0	-1.1080	0.5592	0.2338	-1.9485

B	COL1	COL2	COL3	COL4	COL5	COL6	COL7	COL8	COL9
ROW1	1.0000	0	0	0	0	0	0	0	0
ROW2	0	1.0000	-0.3820	0.4650	-0.6270	0.6920	-0.3080	0.3820	-0.4880
ROW3	0	-0.3820	1.0000	-0.0230	0.5180	-0.3680	0.6660	0.0230	0.4300
ROW4	0	0.4650	-0.0230	1.0000	-0.1410	0.3110	0.0470	0.6750	-0.1150
ROW5	0	-0.6270	0.5180	-0.1410	1.0000	-0.5270	0.4320	-0.1050	0.6730
ROW6	0	0.6920	-0.3680	0.3110	-0.5270	1.0000	-0.4270	0.4720	-0.7140
ROW7	0	-0.3080	0.6660	0.0470	0.4320	-0.4270	1.0000	0.0830	0.5190
ROW8	0	0.3820	0.0230	0.6750	-0.1050	0.4720	0.0830	1.0000	-0.1950
ROW9	0	-0.4880	0.4300	-0.1150	0.6730	-0.7140	0.5190	-0.1950	1.0000

BSWEPT	COL1 RUN	COL2	COL3	COL4	COL5	COL6	COL7	COL8	COL9
ROW1	-1.0000	0	0	0	0	0	0	0	0
ROW2	0	-3.0847	-0.1939	0.9921	-1.3365	1.7897	0.1160	-0.3366	0.7436
ROW3	0	-0.1939	-2.0659	0.0284	0.5877	0.0477	1.1676	0.0119	-0.1682
ROW4	0	0.9921	0.0284	-2.2001	0.1608	-0.6100	-0.0420	1.4082	-0.0285
ROW5	0	-1.3365	0.5877	0.1608	-2.7179	0.5684	-0.1617	0.1326	1.4583
ROW6	0	1.7897	0.0477	-0.6100	0.5684	-3.8898	-0.4984	1.3004	-1.8649

| ROW7 | 0 | 0.1160 | 1.1676 | -0.0420 | -0.1617 | -0.4984 | -2.1982 | 0.4620 |
| 0.5336 | | | | | | | | |

| ROW8 | 0 | -0.3366 | 0.0119 | 1.4082 | 0.1326 | 1.3004 | 0.4620 | -2.4377 |
| 0.1167 | | | | | | | | |

| ROW9 | 0 | 0.7436 | -0.1682 | -0.0285 | 1.4583 | -1.8649 | 0.5336 | 0.1167 |
| -3.1352 | | | | | | | | |

BCAT21	COL1	COL2	COL3	COL4	COL5
ROW1	0	1.7897	0.0477	-0.6100	0.5684
ROW2	0	0.1160	1.1676	-0.0420	-0.1617
ROW3	0	-0.3366	0.0119	1.4082	0.1326
ROW4	0	0.7436	-0.1682	-0.0285	1.4583

BCAT12	COL1	COL2	COL3	COL4
ROW1	0	0	0	0
ROW2	1.7897	0.1160	-0.3366	0.7436
ROW3	0.0477	1.1676	0.0119	-0.1682
ROW4	-0.6100	-0.0420	1.4082	-0.0285
ROW5	0.5684	-0.1617	0.1326	1.4583

BCAT22	COL1	COL2	COL3	COL4
ROW1	-3.8898	-0.4984	1.3004	-1.8649
ROW2	-0.4984	-2.1982	0.4620	0.5336
ROW3	1.3004	0.4620	-2.4377	0.1167
ROW4	-1.8649	0.5336	0.1167	-3.1352

BCAT	COL1	COL2	COL3	COL4	COL5	COL6	COL7	COL8
COL9								
ROW1	-1.0000	0	0	0	0	0	0	0
0								
ROW2	0	-2.1790	-0.2946	0.8351	-1.1080	1.7897	0.1160	-0.3366
0.7436								
ROW3	0	-0.2946	-1.4032	0.2316	0.5592	0.0477	1.1676	0.0119
-0.1682								
ROW4	0	0.8351	0.2316	-1.3468	0.2338	-0.6100	-0.0420	1.4082
-0.0285								
ROW5	0	-1.1080	0.5592	0.2338	-1.9485	0.5684	-0.1617	0.1326
1.4583								
ROW6	0	1.7897	0.0477	-0.6100	0.5684	-3.8898	-0.4984	1.3004
-1.8649								
ROW7	0	0.1160	1.1676	-0.0420	-0.1617	-0.4984	-2.1982	0.4620
0.5336								
ROW8	0	-0.3366	0.0119	1.4082	0.1326	1.3004	0.4620	-2.4377
0.1167								
ROW9	0	0.7436	-0.1682	-0.0285	1.4583	-1.8649	0.5336	0.1167
-3.1352								

C	COL1	COL2	COL3	COL4	COL5	COL6	COL7	COL8
COL9	COL10	COL11	COL12	COL13				

```
ROW1    1.0000        0        0        0        0        0        0        0
            0        0        0        0        0

ROW2        0    1.0000  -0.3700   0.4420  -0.6560   0.7020  -0.2870   0.4050
       -0.4930   0.7380  -0.3360   0.3620  -0.3580

ROW3        0   -0.3700   1.0000   1.0E-03   0.4710  -0.3160   0.6510   0.0380
        0.3550  -0.2760   0.6300   0.1000   0.3560

ROW4        0    0.4420   1.0E-03   1.0000  -0.1610   0.3030   0.0660   0.6840
       -0.1200   0.2890  -0.0160   0.6530  -0.0840

ROW5        0   -0.6560   0.4710  -0.1610   1.0000  -0.5130   0.3990  -0.1230
        0.6420  -0.5750   0.4100  -0.1410   0.5630

ROW6        0    0.7020  -0.3160   0.3030  -0.5130   1.0000  -0.4120   0.4730
       -0.7010   0.7820  -0.3240   0.3140  -0.3850

ROW7        0   -0.2870   0.6510   0.0660   0.3990  -0.4120   1.0000   0.0840
        0.4910  -0.2870   0.5920   0.0870   0.3150

ROW8        0    0.4050   0.0380   0.6840  -0.1230   0.4730   0.0840   1.0000
       -0.2110   0.3760   0.0710   0.7190  -0.0460

ROW9        0   -0.4930   0.3550  -0.1200   0.6420  -0.7010   0.4910  -0.2110
        1.0000  -0.5860   0.3940  -0.1280   0.6310

ROW10       0    0.7380  -0.2760   0.2890  -0.5750   0.7820  -0.2870   0.3760
       -0.5860   1.0000  -0.3670   0.4080  -0.6010

ROW11       0   -0.3360   0.6300  -0.0160   0.4100  -0.3240   0.5920   0.0710
        0.3940  -0.3670   1.0000   0.0930   0.5150

ROW12       0    0.3620   0.1000   0.6530  -0.1410   0.3140   0.0870   0.7190
       -0.1280   0.4080   0.0930   1.0000  -0.1400

ROW13       0   -0.3580   0.3560  -0.0840   0.5630  -0.3850   0.3150  -0.0460
        0.6310  -0.6010   0.5150  -0.1400   1.0000


CSWEPT   COL1     COL2     COL3     COL4    COL5     COL6     COL7     COL8
         COL9    COL10    COL11    COL12    COL13


ROW1   -1.0000        0        0        0        0        0        0        0
            0        0        0        0        0

ROW2        0   -3.2780  -0.3363   0.8239  -1.4705   1.8470   0.2468  -0.1534
        0.6875   0.3297  -0.0811   0.0717   0.0358

ROW3        0   -0.3363  -1.9885   0.0793   0.4889   0.1604   1.1779  -0.00601
       -0.2314   0.0368   0.3868   0.1596   0.1554

ROW4        0    0.8239   0.0793  -2.1440   0.0563  -0.5397  -0.0490   1.4016
        0.0261  -0.0379  -0.0852   0.2711  -0.0994

ROW5        0   -1.4705   0.4889   0.0563  -2.6711   0.5807  -0.1125   0.2260
        1.3331  -0.1222  .0074522  -0.0944   0.2363

ROW6        0    1.8470   0.1604  -0.5397   0.5807  -3.7523  -0.5876   1.1557
       -1.6819   0.5165  -0.0144  -0.0673   0.0868

ROW7        0    0.2468   1.1779  -0.0490  -0.1125  -0.5876  -2.1855   0.4474
        0.5254   0.0644   0.2508  -0.0707  -0.1205

ROW8        0   -0.1534  -0.00601   1.4016   0.2260   1.1557   0.4474  -2.4478
        0.0236  -.004731   0.1561   0.5339   0.1142

ROW9        0    0.6875  -0.2314   0.0261   1.3331  -1.6819   0.5254   0.0236
       -2.8636  -0.0332   0.1012   0.0440   0.5752

ROW10       0    0.3297   0.0368  -0.0379  -0.1222   0.5165   0.0644  -.004731
       -0.0332   0.3045  -0.0873   0.1239  -0.2312
```

```
ROW11        0  -0.0811   0.3868  -0.0862 .0074522  -0.0144    0.2508    0.1561
      0.1012  -0.0873   0.5205   0.0244   0.1956

ROW12        0   0.0717   0.1596   0.2711  -0.0944  -0.0673   -0.0707    0.5339
      0.0440   0.1239   0.0244   0.4168  -0.1021

ROW13        0   0.0358   0.1554  -0.0894   0.2363   0.0868   -0.1205    0.1142
      0.5752  -0.2312   0.1956  -0.1021   0.5306
```

| CCAT11 | COL1 | COL2 | COL3 | COL4 | COL5 | COL6 | COL7 | COL8 |
COL9								
ROW1	-1.0000	0	0	0	0	0	0	0
0								
ROW2	0	-3.0847	-0.1939	0.9921	-1.3365	1.7897	0.1160	-0.3366
0.7436								
ROW3	0	-0.1939	-2.0659	0.0284	0.5877	0.0477	1.1676	0.0119
-0.1682								
ROW4	0	0.9921	0.0284	-2.2001	0.1608	-0.6100	-0.0420	1.4082
-0.0285								
ROW5	0	-1.3365	0.5877	0.1608	-2.7179	0.5684	-0.1617	0.1326
1.4583								
ROW6	0	1.7897	0.0477	-0.6100	0.5684	-3.8898	-0.4984	1.3004
-1.8649								
ROW7	0	0.1160	1.1676	-0.0420	-0.1617	-0.4984	-2.1982	0.4620
0.5336								
ROW8	0	-0.3366	0.0119	1.4082	0.1326	1.3004	0.4620	-2.4377
0.1167								
ROW9	0	0.7436	-0.1682	-0.0285	1.4583	-1.8649	0.5336	0.1167
-3.1352								

CCAT12	COL1	COL2	COL3	COL4
ROW1	0	0	0	0
ROW2	0.3297	-0.0811	0.0717	0.0358
ROW3	0.0368	0.3868	0.1596	0.1554
ROW4	-0.0379	-0.0862	0.2711	-0.0894
ROW5	-0.1222	.0074522	-0.0944	0.2363
ROW6	0.5165	-0.0144	-0.0673	0.0868
ROW7	0.0644	0.2508	-0.0707	-0.1205
ROW8	-.004731	0.1561	0.5339	0.1142
ROW9	-0.0332	0.1012	0.0440	0.5752

| CCAT21 | COL1 | COL2 | COL3 | COL4 | COL5 | COL6 | COL7 | COL8 |
COL9								
ROW1	0	0.3297	0.0368	-0.0379	-0.1222	0.5165	0.0644	-.004731
-0.0332								
ROW2	0	-0.0811	0.3868	-0.0862	.0074522	-0.0144	0.2508	0.1561
0.1012								
ROW3	0	0.0717	0.1596	0.2711	-0.0944	-0.0673	-0.0707	0.5339
0.0440								
ROW4	0	0.0358	0.1554	-0.0894	0.2363	0.0868	-0.1205	0.1142
0.5752								

```
CCAT22    COL1    COL2    COL3    COL4

ROW1     0.3045  -0.0873  0.1239  -0.2312
ROW2    -0.0873   0.5205  0.0244   0.1956
ROW3     0.1239   0.0244  0.4168  -0.1021
ROW4    -0.2312   0.1956 -0.1021   0.5306
```

```
CCAT      COL1    COL2    COL3    COL4    COL5    COL6    COL7    COL8
    COL9    COL10   COL11   COL12   COL13

ROW1   -1.0000        0       0       0       0       0       0       0
       0       0       0       0       0

ROW2        0  -3.0847 -0.1939  0.9921 -1.3365  1.7897  0.1160 -0.3366
  0.7436  0.3297 -0.0811  0.0717  0.0358

ROW3        0  -0.1939 -2.0659  0.0284  0.5877  0.0477  1.1676  0.0119
 -0.1682  0.0368  0.3868  0.1596  0.1554

ROW4        0   0.9921  0.0284 -2.2001  0.1608 -0.6100 -0.0420  1.4082
 -0.0285 -0.0379 -0.0862  0.2711 -0.0894

ROW5        0  -1.3365  0.5877  0.1608 -2.7179  0.5684 -0.1617  0.1326
  1.4583 -0.1222 .0074522 -0.0944  0.2363

ROW6        0   1.7897  0.0477 -0.6100  0.5684 -3.8898 -0.4984  1.3004
 -1.8649  0.5165 -0.0144 -0.0673  0.0868

ROW7        0   0.1160  1.1676 -0.0420 -0.1617 -0.4984 -2.1982  0.4620
  0.5336  0.0644  0.2508 -0.0707 -0.1205

ROW8        0  -0.3366  0.0119  1.4082  0.1326  1.3004  0.4620 -2.4377
  0.1167 -.004731  0.1561  0.5339  0.1142

ROW9        0   0.7436 -0.1682 -0.0285  1.4583 -1.8649  0.5336  0.1167
 -3.1352 -0.0332  0.1012  0.0440  0.5752

ROW10       0   0.3297  0.0368 -0.0379 -0.1222  0.5165  0.0644 -.004731
 -0.0332  0.3045 -0.0873  0.1239 -0.2312

ROW11       0  -0.0811  0.3868 -0.0862 .0074522 -0.0144  0.2508  0.1561
  0.1012 -0.0873  0.5205  0.0244  0.1956

ROW12       0   0.0717  0.1596  0.2711 -0.0944 -0.0673 -0.0707  0.5339
  0.0440  0.1239  0.0244  0.4168 -0.1021

ROW13       0   0.0358  0.1554 -0.0894  0.2363  0.0868 -0.1205  0.1142
  0.5752 -0.2312  0.1956 -0.1021  0.5306
```

```
MAXEST    COL1    COL2    COL3    COL4    COL5    COL6    COL7    COL8
COL9    COL10   COL11   COL12   COL13

ROW1   -1.0000        0       0       0       0       0       0       0
   0       0       0       0       0

ROW2        0   1.0000 -0.3820  0.4650 -0.6270  0.6920 -0.3080  0.3820
 -0.4880  0.7266 -0.3506  0.3537 -0.3532

ROW3        0  -0.3820  1.0000 -0.0230  0.5180 -0.3680  0.6660  0.0230
  0.4300 -0.3131  0.6432  0.0859  0.4039

ROW4        0   0.4650 -0.0230  1.0000 -0.1410  0.3110  0.0470  0.6750
 -0.1150  0.2960 -0.0328  0.6452 -0.0773

ROW5        0  -0.6270  0.5180 -0.1410  1.0000 -0.5270  0.4320 -0.1050
  0.6730 -0.5707  0.4385 -0.1164  0.5842
```

ROW6	0	0.6920	-0.3680	0.3110	-0.5270	1.0000	-0.4270	0.4720
	-0.7140	0.7777	-0.3493	0.3084	-0.4032			
ROW7	0	-0.3080	0.6660	0.0470	0.4320	-0.4270	1.0000	0.0830
	0.5190	-0.3054	0.6043	0.0813	0.3408			
ROW8	0	0.3820	0.0230	0.6750	-0.1050	0.4720	0.0830	1.0000
	-0.1950	0.3649	0.0694	0.7117	-0.0349			
ROW9	0	-0.4880	0.4300	-0.1150	0.6730	-0.7140	0.5190	-0.1950
	1.0000	-0.5906	0.4321	-0.1098	0.6470			
ROW10	0	0.7266	-0.3131	0.2960	-0.5707	0.7777	-0.3054	0.3649
	-0.5906	0.9909	-0.3877	0.3982	-0.6088			
ROW11	0	-0.3506	0.6432	-0.0328	0.4385	-0.3493	0.6043	0.0694
	0.4321	-0.3877	1.0150	0.0884	0.5428			
ROW12	0	0.3537	0.0859	0.6452	-0.1164	0.3084	0.0813	0.7117
	-0.1098	0.3982	0.0884	0.9904	-0.1261			
ROW13	0	-0.3532	0.4039	-0.0773	0.5842	-0.4032	0.3408	-0.0349
	0.6470	-0.6088	0.5428	-0.1261	1.0179			

To check whether the estimates are reasonable, they can be compared with those using a listwise deletion matrix and a pairwise deletion matrix generated for the same set of data. As these matrices represent extremes (listwise=minimum data that can be used; pairwise=maximum data that can be used), the maximum-likelihood estimates should fall somewhere in between.

The panel estimates are, in general, somewhat smaller than the estimates generated assuming the data were cross-sectional. Smaller correlation tends to decrease the likelihood of multicollinearity in the data.

4. Discussion

The estimation of missing data in longitidinal data sets can be thought of as a two-stage process. First, an adequate description of the pattern and degree of missing data should be generated. A good description of the missingness will help the researcher to interpret any subsequent analyses. Finally, a maximum-likelihood method will generally yield good estimates. In a cross-sectional design, the EM algorithm will suffice for those who have access to BMDP. For longitudinal data, some additional programming in a matrix language may be necessary.

We have concentrated here on the nested pattern of missing data. If only a few participants in a study fail to conform to the nested pattern, they can be eliminated. If, however, a large percentage of participants either enter the study at follow-up waves or drop out and come back into the study, other estimation procedures may be necessary (Little & Rubin, 1987). Such procedures are similar to those presented here but tend to be somewhat more difficult to compute.

Chapter 2

Alternative Analyses of Repeated Measures Designs by ANOVA and MANOVA

1. Summary of Method

Analysis of repeated measures data can be accomplished using a number of different methods (Games, 1990). While some of the alternatives represent different algorithms generating the same result, others represent tests of qualitatively different hypotheses. The methods exemplified here show the univariate and multivariate alternatives for the computation of the repeated measures ANOVA (Bock, 1975; McCall and Appelbaum, 1973; Hertzog and Rovine, 1985).

First, the univariate solution to the analysis of variance for a single repeatedly measured dependent variable will be demonstrated. The omnibus test will be computed for an analysis of variance with one grouping factor and one repeated measures factor.

The univariate solution attempts to discern differences in means for a repeatedly measured variable. Additional categorical grouping variables and interval level covariates can be included in the analysis. The univariate repeated measures ANOVA differs from the the cross-sectional ANOVA in at least three ways. Each individual will appear in more than one cell. Each cell consists of only one measure. The error term used is the interaction between an identification factor and the repeated measures factor. A simple repeated measures model with six subjects each measured three times on variable **Y** (Y_1, Y_2, Y_3) would look like

			TIME	
		T_1	T_2	T_3
SUBJ	1	20	30	40
	2	23	31	41
	3	18	21	39
	4	20	30	40
	5	10	18	50
	6	24	38	50.

Each data point, **Y**, would be identified by two values, **TIME** and **SUBJ**. The model is univariate in the sense that if **TIME** and **SUBJ** are used as factors, the dependent variable is defined as **Y**. (not Y_1, Y_2, and Y_3). The model for the ANOVA would include three effects: **TIME**, **SUBJ**, and **TIME*SUBJ** with the last term functioning as the error term for the analysis.

Next, the MANOVA solution to the univariate problem presented in the above example will be demonstrated. The **n** occasions of the dependent variable are first transformed into a new set representing polynomial trends. This would create the following set of data:

		TRENDS	
		T_{lin}	T_{quad}
SUBJ	1	20	0
	2	18	2
	3	21	15
	4	20	0
	5	40	24
	6	26	-4.

These trend scores were creating using orthogonal polynomial coefficients for linear and quadratic trends (McCall and Appelbaum, 1973). The model to analyze these data would be multivariate in that two dependent variables, Y_{lin} and Y_{quad}, would be analyzed. There is now no independent **TIME** variable as that effect has been incorporated into the dependent trend variables. A multivariate test of the transformed variables represents an alternative to the omnibus test mentioned above.

The univariate test, which is generally more powerful than the multivariate test, requires a certain pattern of covariation among the repeated measures. If this pattern does not hold, the univariate strategy may not be appropriate for a given data set (Games, 1990). The test of this structure has been termed sphericity (Mauchly, 1940; Anderson, 1958). A variance/covariance matrix of the transformed variables can be used to test sphericity and thus the appropriateness of the univariate F-test.

In addition to generating an alternative F-test, the decomposition of the repeated measures into components can be used to test specific patterns in the means according to the logic of planned contrasts (Hertzog and Rovine, 1985).

The selection of follow-up strategies is often difficult (Games, 1990). We will suggest a general form for a test of a follow-up contrast that can be computed using a contrast matrix and a variance/covariance matrix of the dependent variable scores.

2. Outline of the Computational Procedures

With its most recent release (SAS Statistics, 1985), SAS PROC GLM allows great flexibility for computing the different repeated measures models. In particular, the REPEATED option allows the user to estimate both the univariate and multivariate models. It includes tests of sphericity and has options for computing planned contrasts and certain follow-up tests. Specifications for the different models are, however, occasionally confusing. We will first use PROC GLM to present the univariate solution to a repeated measures problem and then to contrast that with the multivariate solution to the same problem.

3. An Annotated Data Example

Step 1: Data Input

This uses the data listed above and represents a simple model with six cases of data and three repeated measures of a dependent variable (DEPT1 DEPT2 DEPT3) representing three times of measurement. Participants in this imagined study were assigned to one of two groups (GROUP) with subjects 1 to 3 in Group 1 and subjects 4 to 6 in Group 2. The groups represent an imagined experimental treatment and a control. This example is kept simple to allow us to show how the different repeated measures setups yield the same results. The SAS input statement would appear as follows:

```
INPUT ID GROUP TIME DEP;
```

Each line of data represents a "case" with a set of coding vectors that uniquely define the value of the dependent variable DEPT. With six subjects this setup has 18 lines of data: this represents an N of three times the number of subjects. This replicated N represents the source of the power of the univariate F-test.

Most data sets, however, are not set up in this fashion. Normally, all occasions of the repeated measure appear on the same case. An SAS data step to read the raw data would look as follows:

```
INPUT ID    GROUP    TIME DEPT1
#2    ID2   GROUP2   TIME2 DEPT2
#3    ID3   GROUP3   TIME3 DEPT3;
```

In this setup, each case includes a complete set of repeated dependent variables. To run the analysis using SAS, one has traditionally been forced to rewrite the data set to identify each observation with a set of coding vectors including one vector representing the effect TIME. The REPEATED option in PROC GLM essentially performs this step thus tripling the number of cases in the data set and adding in the appropriate coding vectors.

Step 2:   The Programs with Annotated Outputs

The Univariate Approach.   The PROC GLM setup for the first input statement shown above is:

```
OPTIONS NOCENTER LINESIZE=80 NONUMBER NODATE;
DATA;
INFILE CARDS;
INPUT ID GROUP TIME DEP;
CARDS;
1    1   1   20
1    1   2   30
1    1   3   40
2    1   1   23
2    1   2   31
2    1   3   41
3    1   1   18
3    1   2   21
3    1   3   39
4    2   1   20
4    2   2   30
4    2   3   40
5    2   1   10
5    2   2   18
5    2   3   50
6    2   1   24
6    2   2   38
6    2   3   50
;
PROC GLM;
 CLASSES ID GROUP TIME;
 MODEL DEP=GROUP ID(GROUP) TIME TIME*GROUP;
 MEANS GROUP TIME TIME*GROUP;
 LSMEANS GROUP TIME TIME*GROUP;
 CONTRAST 'LINEAR'
    TIME -1 0 1;
 TEST H=GROUP
      E=ID(GROUP);
```

TIME and ID are the two coding vectors that uniquely determine each value of the dependent variable by associating that value with a subject and an occasion.

The results of this analysis follow:

```
GENERAL LINEAR MODELS PROCEDURE

CLASS LEVEL INFORMATION

CLASS      LEVELS     VALUES

ID           6        1 2 3 4 5 6

GROUP        2        1 2

TIME         3        1 2 3


NUMBER OF OBSERVATIONS IN DATA SET = 18

GENERAL LINEAR MODELS PROCEDURE

DEPENDENT VARIABLE: DEP
```

SOURCE	DF	SUM OF SQUARES	MEAN SQUARE	F VALUE
MODEL	9	2120.94444444	235.66049383	9.45
ERROR	8	199.55555556	24.94444444	PR > F
CORRECTED TOTAL	17	2320.50000000		0.002

F-SQUARE	C.V.	ROOT MSE	DEP MEAN
0.914003	16.5562	4.99444135	30.16666667

SOURCE	DF	TYPE I SS	F VALUE	PR > F
GROUP	1	16.05555556	0.64	0.4456
ID(GROUP)	4	249.11111111	2.50	0.1260
TIME	2	1794.33333333	35.97	0.0001
GROUP*TIME	2	61.44444444	1.23	0.3417

SOURCE	DF	TYPE III SS	F VALUE	PR > F
GROUP	1	16.05555556	0.64	0.4456
ID(GROUP)	4	249.11111111	2.50	0.1260
TIME	2	1794.33333333	35.97	0.0001
GROUP*TIME	2	61.44444444	1.23	0.3417

```
TESTS OF HYPOTHESES USING THE TYPE III MS FOR ID(GROUP) AS AN ERROR TERM
```

SOURCE	DF	TYPE III SS	F VALUE	PR > F
GROUP	1	16.05555556	0.26	0.6383

```
MEANS
```

GROUP	N	DEP
1	9	29.2222222
2	9	31.1111111

```
TIME         N          DEP

1            6     19.1666667
2            6     28.0000000
3            6     43.3333333


GROUP   TIME        N           DEP

1       1           3     20.3333333
1       2           3     27.3333333
1       3           3     40.0000000
2       1           3     18.0000000
2       2           3     28.6666667
2       3           3     46.6666667


LEAST SQUARES MEANS

GROUP            DEP
               LSMEAN

1          29.2222222
2          31.1111111


TIME             DEP
               LSMEAN

1          19.1666667
2          28.0000000
3          43.3333333


GROUP   TIME          DEP
                    LSMEAN

1       1       20.3333333
1       2       27.3333333
1       3       40.0000000
2       1       18.0000000
2       2       28.6666667
2       3       46.6666667
```

The effects of the analysis are GROUP, TIME, and GROUP*TIME. The error term has been partitioned into two components: ID(GROUP) and ERROR. ID(GROUP) is the error associated with the between effect, and ERROR is associated with the within effect. Since all effects in the model are tested against the residual (ERROR), a special test for GROUP must be requested using a TEST statement. TYPE III sums of squares representing the independent contribution for each effect are appropriate for testing each hypothesis related to an effect.

The REPEATED option of PROC GLM now allows one to avoid rewriting the data set. The following code will generate the same solution:

```
OPTIONS NOCENTER LINESIZE=80 NONUMBER NODATE;
DATA;
INFILE CARDS;
INPUT ID    GROUP    TIME DEPT1
#2    ID2   GROUP2   TIME2 DEPT2
#3    ID3   GROUP3   TIME3 DEPT3;
```

```
CARDS;
<DATA>
;
PROC GLM;
  CLASSES GROUP;
  MODEL DEPT1 DEPT2 DEPT3=GROUP;
  MEANS GROUP;
  LSMEANS GROUP;
  REPEATED TIME 3 POLYNOMIAL/SUMMARY;
```

The REPEATED option creates a variable called OBS which functions as did ID in the example above.

The solution obtained from this setup appears below. As can be seen, the results match those of the rewritten data set.

GENERAL LINEAR MODELS PROCEDURE

TESTS OF HYPOTHESES FOR BETWEEN SUBJECTS EFFECTS

SOURCE	DF	TYPE III SS	MEAN SQUARE	F VALUE	PR > F
GROUP	1	16.05555556	16.05555556	0.26	0.6383
ERROR	4	249.11111111	62.27777778		

GENERAL LINEAR MODELS PROCEDURE

UNIVARIATE TESTS OF HYPOTHESES FOR WITHIN SUBJECT EFFECTS

SOURCE: TIME

DF	TYPE III SS	MEAN SQUARE	F VALUE	PR > F	ADJ PR > F G - G	H -
2	1794.33333333	897.16666667	35.97	0.0001	0.0022	0.000

SOURCE: TIME*GROUP

DF	TYPE III SS	MEAN SQUARE	F VALUE	PR > F	ADJ PR > F G - G	H -
2	61.44444444	30.72222222	1.23	0.3417	0.3332	0.340

SOURCE: ERROR(TIME)

DF	TYPE III SS	MEAN SQUARE
8	199.55555556	24.94444444

GREENHOUSE-GEISSER EPSILON = 0.5781
    HUYNH-FELDT EPSILON = 0.868`

Three tests are produced for each effect including a repeated measures factor. In addition to the significance level of the omnibus F-test, two probability levels adjusted for violation of the sphericity assumption are included: the Greenhouse–Geisser and the Huynh–Feldt. Both compute a significance level for the value of F distributed over the adjusted degrees of freedom. Games (1990) discusses the selection of the most appropriate significance level.

Although this approach contains multiple dependent variables, it is nevertheless a univariate analysis. With the levels of the time effect locating the values of the dependent variable, only one variable (for each set of repeated measures) appears on the dependent side of the model.

The Multivariate Approach.   An alternative approach to the above analysis involves transforming the repeated measured variables into a set of equivalent dependent variables using some orthogonal set of transformations coefficients, often orthonormalized polynomial coefficients. If, for example, one started with a variable measured on three occasions, the original three variables would be transformed into three new variables representing a mean level and two orthogonal (linear and quadratic) polynomial contrasts.

The following SAS statement can be used to transform the data:

```
OPTIONS NOCENTER LINESIZE=80 NONUMBER NODATE; DATA; INFILE CARDS; INPUT ID   GROUP
TIME DEPT1 #2    ID2  GROUP2  TIME2 DEPT2 #3    ID3  GROUP3  TIME3 DEPT3;

* COEFFICIENTS FOR EVEN SPACING;

ME1=1/SQRT(3); ME2=1/SQRT(3); ME3=1/SQRT(3);
LE1=-1/SQRT(2);    LE2= 0/SQRT(2);    LE3=1/SQRT(2);
QE1=1/SQRT(6);    QE2=-2/SQRT(6);    QE3=1/SQRT(6);

DEPTMEAN=ME1*DEPT1+ME2*DEPT2+ME3*DEPT3;
DEPTLIN=LE1*DEPT1+LE2*DEPT2+LE3*DEPT3;
DEPTQUAD=QE1*DEPT1+QE2*DEPT2+QE3*DEPT3;

CONSTANT=1;

CARDS;
<DATA>
;
PROC GLM;
 CLASSES GROUP;
 MODEL DEPTMEAN=GROUP;
 MEANS GROUP;
 LSMEANS GROUP;
PROC GLM;
 CLASSES GROUP;
 MODEL DEPTLIN DEPTQUAD=CONSTANT GROUP/NOINT;
PROC CORR;
 VAR DEPTLIN; WITH DEPTQUAD;
```

Appropriate coefficients for more than three occasions along with a method for computing coefficients for unequal spacing appear in Myers (1979). SAS PROC GLM can, however, be used to generate the coefficients for a number of different transformations.

Tests of the linear and quadratic effects can be computed by regressing the trend scores onto a constant. With the addition of a grouping variable, one can test whether the linear (or quadratic) trend differs according to group membership. Between-group effects can be tested by using the transformed mean level variable. This run generated the following test:

```
GENERAL LINEAR MODELS PROCEDURE

DEPENDENT VARIABLE: DEPTMEAN
```

SOURCE	DF	SUM OF SQUARES	MEAN SQUARE	F VALU
MODEL	1	16.05555556	16.05555556	0.26
ERROR	4	249.11111111	62.27777778	PR > F
CORRECTED TOTAL	5	265.16666667		0.638

R-SQUARE	C.V.	ROOT MSE	DEPTMEAN MEAN
0.060549	15.1035	7.89162707	52.25019936

SOURCE	DF	TYPE I SS	F VALUE	PR > F
GROUP	1	16.05555556	0.26	0.6383

SOURCE	DF	TYPE III SS	F VALUE	PR > F
GROUP	1	16.05555556	0.26	0.6383

MEANS

GROUP	N	DEPTMEAN
1	3	50.6143736
2	3	53.8860251

LEAST SQUARES MEANS

GROUP	DEPTMEAN LSMEAN
1	50.6143736
2	53.8860251

CLASS LEVEL INFORMATION

CLASS	LEVELS	VALUES
GROUP	2	1 2

NUMBER OF OBSERVATIONS IN DATA SET = 6

DEPENDENT VARIABLE: DEPTLIN

SOURCE	DF	SUM OF SQUARES	MEAN SQUARE	F VALUE
MODEL	2	1812.83333333	906.41666667	33.67
ERROR	4	107.66666667	26.91666667	PR > F
UNCORRECTED TOTAL	6	1920.50000000		0.003

R-SQUARE	C.V.	ROOT MSE	DEPTLIN MEAN
0.943938	30.3605	5.18812747	17.08841388

SOURCE	DF	TYPE I SS	F VALUE	PR > F
CONSTANT	1	1752.08333333	65.09	0.0013
GROUP	1	60.75000000	2.26	0.2074

SOURCE	DF	TYPE III SS	F VALUE	PR > F
CONSTANT	0	0.00000000	.	.
GROUP	1	60.75000000	2.26	0.2074

```
DEPENDENT VARIABLE: DEPTQUAD
```

SOURCE	DF	SUM OF SQUARES	MEAN SQUARE	F VALUE
MODEL	2	42.94444444	21.47222222	0.93
ERROR	4	91.88888889	22.97222222	PR >
UNCORRECTED TOTAL	6	134.83333333		0.464

R-SQUARE	C.V.	ROOT MSE	DEPTQUAD MEAN
0.318500	180.6191	4.79293461	2.65361389

SOURCE	DF	TYPE I SS	F VALUE	PR > F
CONSTANT	1	42.25000000	1.84	0.2465
GROUP	1	0.69444444	0.03	0.8704

SOURCE	DF	TYPE III SS	F VALUE	PR > F
CONSTANT	0	0.00000000	.	.
GROUP	1	0.69444444	0.03	0.8704

VARIABLE	N	MEAN	STD DEV	SUM	MINIMUM	MAXIMUM
DEPTLIN	6	17.08841	5.803734	102.5305	12.72792	28.28427
DEPTQUAD	6	2.65361	4.303100	15.9217	-0.81650	9.79796

```
PEARSON CORRELATION COEFFICIENTS / PROB > |R| UNDER HO:RHO=0 / N = 6

          DEPTLIN

DEPTQUAD  0.73168
          0.0983
```

To interpret the output, it is necessary to read "CONSTANT" as "TIME". Each within effect is actually the effect *TIME(linear) or *TIME(quadratic) interaction. So, for example, the effect CONSTANT is actually the CONSTANT*TIME interaction or the TIME effect. This strategy generates separate linear and quadratic TIME effects. It also partitions the error term in separate terms appropriate for testing the separate trends. Comparing this output with the univariate solution shows how the two strategies generate the same ANOVA. Summing the sums of squares for the linear and quadratic effects of the multivariate solution yields the omnibus sums of squares for the univariate TIME effect. Summing the separate error terms for both effects yields the omnibus error term. The transformations serve to decompose the sums of squares into linear and quadratic components. Note that the test of CONSTANT appears as a TYPE I SS test while all other hypotheses should be tested using TYPE III SS. In the regression sense, CONSTANT is predicting none of variance of the dependent variable above and beyond that predicted by the "real" predictors in the equation.

If the trends are of interest, the univariate tests of, in this case, the LINEAR and QUADRATIC trends suffice. If one is, however, interested in an overall TIME effect, one must at this point select a test.

In addition to the ANOVA, the variance/covariance matrix of the tranformed component scores (in this case the linear and quadratic components) can be computed. This off-diagonal element can indicate sphericity with a significant correlation representing violation of the assumption. If sphericity has been violated, the pooled error term (linear and quadratic) may yield an inappropriate F-test.

If violation of sphericity is a concern, an alternate strategy would compute the "true" multivariate test of the linear and quadratic effects (Games, 1990; Hertzog and Rovine, 1985).

The F-test for this approach has less power than the corresponding univariate test; however, the assumption of sphericity is unneccessary for this analysis. It represents a viable computational alternative.

The "true" multivariate test can be obtained either by including a MANOVA statement in the PROC that tests the Linear and Quadratic effects or by using the REPEATED option along with the specification of POLYNOMIAL as the transformation. (Note: The POLYNOMIAL option allows one to compute coefficients for unequally spaced variables.) Some of the output for the latter option follows:

```
OPTIONS NOCENTER LINESIZE=80 NONUMBER NODATE;
DATA;
INFILE CARDS;
INPUT ID    GROUP    TIME DEPT1
#2     ID2   GROUP2   TIME2 DEPT2
#3     ID3   GROUP3   TIME3 DEPT3;

CARDS;
<DATA>
;
PROC GLM;
  CLASSES GROUP;
  MODEL DEPT1 DEPT2 DEPT3=GROUP;
  MEANS GROUP;
  LSMEANS GROUP;
  REPEATED TIME 3 POLYNOMIAL/SUMMARY;


GENERAL LINEAR MODELS PROCEDURE


TESTS OF HYPOTHESES FOR BETWEEN SUBJECTS EFFECTS
```

SOURCE	DF	TYPE III SS	MEAN SQUARE	F VALUE	PR > F
GROUP	1	16.05555556	16.05555556	0.26	0.6383
ERROR	4	249.11111111	62.27777778		

```
UNIVARIATE TESTS OF HYPOTHESES FOR WITHIN SUBJECT EFFECTS

SOURCE: TIME
                                                       ADJ  PR > F
     DF      TYPE III SS      MEAN SQUARE   F VALUE  PR > F  G - G   H -
      2    1794.33333333     897.16666667     35.97  0.0001  0.0022  0.000

SOURCE: TIME*GROUP
                                                       ADJ  PR > F
     DF      TYPE III SS      MEAN SQUARE   F VALUE  PR > F  G - G   H -
      2      61.44444444      30.72222222      1.23  0.3417  0.3332  0.340

SOURCE: ERROR(TIME)

     DF      TYPE III SS      MEAN SQUARE
      8     199.55555556      24.94444444

GREENHOUSE-GEISSER EPSILON = 0.5781
       HUYNH-FELDT EPSILON = 0.8681

GENERAL LINEAR MODELS PROCEDURE

REPEATED MEASURES ANALYSIS OF VARIANCE

REPEATED MEASURES LEVEL INFORMATION

DEPENDENT VARIABLE       DEPT1    DEPT2    DEPT3

    LEVEL OF TIME           1        2        3


GENERAL LINEAR MODELS PROCEDURE

MANOVA TEST CRITERIA FOR THE HYPOTHESIS OF NO TIME EFFECT

H = TYPE III SS&CP MATRIX FOR:   TIME
E = ERROR SS&CP MATRIX
P = DF OF RM EFFECT  =      2
Q = HYPOTHESIS DF    =      1
NE= DF OF E          =      4
S = MIN(P,Q)         =      1
M = .5(ABS(P-Q)-1)   =    0.0
N = .5(NE-P-1)       =    0.5
-----------------------------------------------------------------



WILKS' CRITERION       L = DET(E)/DET(H+E) =      0.02204250

        EXACT F = (1-L)/L*(NE+Q-P)/P
                        WITH P AND NE+Q-P DF

            F(2,3) =     66.55     PROB > F = 0.0033
-----------------------------------------------------------------
PILLAI'S TRACE         V = TR(H*INV(H+E)) =      0.97795750

        F APPROXIMATION = (2N+S+1)/(2M+S+1) * V/(S-V)
                        WITH S(2M+S+1) AND S(2N+S) DF

            F(2,3) =     66.55     PROB > F = 0.0033
-----------------------------------------------------------------
HOTELLING-LAWLEY TRACE = TR(E**-1*H) =         44.36690529

        F APPROXIMATION = 2(S*N+1)*TR(E**-1*H)/(S*S*(2M+S+1))
                        WITH S(2M+S+1) AND 2(S*N+1) DF

            F(2,3) =     66.55     PROB > F = 0.0033
-----------------------------------------------------------------
```

```
ROY'S MAXIMUM ROOT CRITERION =                    44.36690529

         FIRST CANONICAL VARIABLE YIELDS AN F UPPER BOUND

              F(2,3) =    66.55     PROB > F = 0.0033

-------------------------------------------------------------


MANOVA TEST CRITERIA FOR THE HYPOTHESIS OF NO TIME*GROUP EFFECT

H = TYPE III SS&CP MATRIX FOR:  TIME*GROUP
E = ERROR SS&CP MATRIX
P = DF OF RM EFFECT    =        2
Q = HYPOTHESIS DF      =        1
NE= DF OF E            =        4
S = MIN(P,Q)           =        1
M = .5(ABS(P-Q)-1)     =      0.0
N = .5(NE-P-1)         =      0.5

-------------------------------------------------------------


WILKS' CRITERION       L = DET(E)/DET(H+E) =      0.37134736

        EXACT F = (1-L)/L*(NE+Q-P)/P
                     WITH P AND NE+Q-P DF

              F(2,3) =     2.54     PROB > F = 0.2263

-------------------------------------------------------------

PILLAI'S TRACE         V = TR(H*INV(H+E)) =       0.62865264

        F APPROXIMATION = (2N+S+1)/(2M+S+1) * V/(S-V)
                     WITH S(2M+S+1) AND S(2N+S) DF

              F(2,3) =     2.54     PROB > F = 0.2263

-------------------------------------------------------------

HOTELLING-LAWLEY TRACE = TR(E**-1*H) =            1.69289648

        F APPROXIMATION = 2(S*N+1)*TR(E**-1*H)/(S*S*(2M+S+1))
                     WITH S(2M+S+1) AND 2(S*N+1) DF

              F(2,3) =     2.54     PROB > F = 0.2263

-------------------------------------------------------------

ROY'S MAXIMUM ROOT CRITERION =                    1.69289648

         FIRST CANONICAL VARIABLE YIELDS AN F UPPER BOUND

              F(2,3) =     2.54     PROB > F = 0.2263

-------------------------------------------------------------


ANALYSIS OF VARIANCE OF CONTRAST VARIABLES

TIME.N REPRESENTS THE NTH DEGREE POLYNOMIAL CONTRAST FOR TIME

CONTRAST VARIABLE: TIME.1

SOURCE            DF    TYPE III SS    MEAN SQUARE    F VALUE    PR > F

MEAN               1   1752.0833333   1752.0833333     65.09    0.0013
GROUP              1     60.7500000     60.7500000      2.26    0.2074

ERROR              4    107.6666667     26.9166667
```

```
CONTRAST VARIABLE: TIME.2
```

SOURCE	DF	TYPE III SS	MEAN SQUARE	F VALUE	PR > F
MEAN	1	42.25000000	42.25000000	1.84	0.2465
GROUP	1	0.69444444	0.69444444	0.03	0.8704
ERROR	4	91.88888889	22.97222222		

This setup generates both the univariate and multivariate solutions for the ANOVA along with tests of sphericity. With this option, the user is given everything and the kitchen sink and can then select the most appropriate analysis (see Hertzog and Rovine, 1985). Although the different "true" multivariate tests (Wilks' λ; Hotelling–Lawley trace; Pilli's trace, and Roy's maximum root) yield the same F-test value for this model, they can represent tests of different hypotheses. Morrison (1976) discusses the differences among these four tests.

When hypotheses exist concerning a particular subset of means, planned contrasts can be used in lieu of the omnibus test to assess the hypothesis. The SAS contrast statement readily computes these for both between and within effects.

The selection of an appropriate follow-up test often presents the researcher with a difficult decision (see Games, 1990). With a major caveat for the analyst to beware, we present a rather general method providing a test of follow-up contrasts (Hertzog and Rovine, 1985, technical supplement; Mitzel and Games, 1981; Keselman, 1982).

The general form of the contrast is

$$t_i = PSI_i / [V(PSI_i)^{1/2}] \qquad (1)$$

where the numerator, $\mathbf{PSI}_i$, is the matrix product of a set of weights and set of means, and the denominator represents the standard error computed specifically for the contrast (Mitzel and Games, 1981). Suppose we wish to test whether the linear effect differs for each group. The transpose of the contrast vector for this hypothesis would be

$$c' = [-1 \quad 0 \quad 1 \quad 1 \quad 0 \quad -1] \qquad (2)$$

and

$$PSI_i = c' * \overline{Y} \qquad (3)$$

where $\overline{Y}$ is the vector of means for group 1 and group 2. The first three coefficients test the linear effect for group 1 and the second three for group 2. The signs of the group 2 coefficients represent our attempt to test whether the linear effect differs for the two groups.

The standard error is computed by

$$V_{lin1-lin2} = c'* S * c/n \tag{4}$$

where n is the subgroup sample size. The *harmonic n* can be used if the subgroups sample sizes differ. S is a variance/covariance matrix for groups 1 and 2 that takes the following form:

$$S = \begin{bmatrix} S_{group1} & 0 \\ 0 & S_{group2} \end{bmatrix} \tag{5}$$

A standard error can be computed by applying a set of weights to a pair of subgroup variance/covariance matrices. The SAS PROC IML setup follows:

```
OPTIONS NOCENTER LINESIZE=80 NONUMBER NODATE;
PROC IML;
START;
COVGRP1= {6.33333 11.8333 2.5,
   11.8333 30.3333 5.0,
   2.5 5.0 1.0};
MEANGRP1= {20.33333, 27.33333, 40.00000};
COVGRP2= {52.0 72.0 -10.0,
   72.0 101.333 -6.66667,
-10.0 -6.66667 33.3333};
MEANGRP2={18.00000, 28.66667, 46.66667};
MEANS=MEANGRP1//MEANGRP2;
ZERO={0 0 0, 0 0 0, 0 0 0};
COVTOP = COVGRP1||ZERO;
COVBOT=ZERO||COVGRP2;
COV=COVTOP//COVBOT;
N=3;
C={-1, 0, 1, 1, 0, -1};
PSI= C`*MEANS;
V=C`*COV*C;
V=V/N;
V=SQRT(V);
T=PSI/V;
PRINT MEANS;
PRINT COV;
PRINT C;
PRINT PSI;
PRINT V;
PRINT T;
FINISH;
RUN;
```

This program yields the following solution:

MEANS	COL1
ROW1	20.3333
ROW2	27.3333
ROW3	40.0000
ROW4	18.0000
ROW5	28.6667
ROW6	46.6667

COV	COL1	COL2	COL3	COL4	COL5	COL6
ROW1	6.3333	11.8333	2.5000	0	0	0
ROW2	11.8333	30.3333	5.0000	0	0	0
ROW3	2.5000	5.0000	1.0000	0	0	0
ROW4	0	0	0	52.0000	72.0000	-10.0000
ROW5	0	0	0	72.0000	101.3	-6.6667
ROW6	0	0	0	-10.0000	-6.6667	33.3333

C	COL1
ROW1	-1.0000
ROW2	0
ROW3	1.0000
ROW4	1.0000
ROW5	0
ROW6	-1.0000

PSI	COL1
ROW1	-9.0000

V	COL1
ROW1	5.9907

T	COL1
ROW1	-1.5023

The *alpha* level for significance should be set to reflect the number of contrasts performed.

4. Discussion

We have presented examples of how to compute both the univariate and multivariate solutions to the repeated measures ANOVA problem. The selection of the most appropriate analysis will depend on a number of issues. As mentioned, the selection of the univariate strategy requires a particular pattern of covariation for the dependent variables across time.

Selection of a strategy may also depend on what specific questions are to be addressed. If one's study attempts to a particular kind of change, one could test a hypothesis regarding that kind of change without recourse to the omnibus test. With multiple occasions only certain trends may make sense. Testing only those trends would represent the selection of a certain set of planned comparisons with degrees of freedom for the set of tests less than the degrees of freedom of the omnibus test. Such a set of planned comparisons could also avoid the problem that can be caused when a particular kind of trend is masked by a number of non-significant effects included to generate the omnibus test.

Chapter 3

Longitudinal Factor Analysis

1. Summary of Method

Longitudinal factor analytic models (Tisak and Meredith, 1990; McArdle and Aber, 1990) allow one to make statements about sets of variables measured on multiple occasions. These models can help formulate a basis for talking about change by defining measurement characteristics of latent constructs (Hertzog and Nesselroade, 1987).

Longitudinal factor analysis is often used when one can assume that a latent variable can be captured by its effect on a set of observed measures. When change in that latent variable is of interest and the observed measures are taken at multiple points in time, that change can be described when certain assumptions can be made.

Imagine an analysis in which a latent variable in part determines three indicators (observed variables). Each is measured at three points in time and on samples from two different populations. Imagine also that factor scores will be generated and used to show both change in the latent variable from occasion to occasion and possible differences in populations. As with any measurement, one must be sure that the scale used at each assessment is the same. If not, any change assessed could be the result of either change in the construct or change in the way one measures the construct (or in the worst case, both). The requirement that the factor structure be the same at each occasion for repeatedly measured indicators of latent constructs is a type of factor invariance that has been termed stationarity (Tisak and Meredith, 1990). If multiple populations are to be considered and one wishes to explore possible differences in populations, once again, it is essential that the same measure be used in each population. An additional condition for invariance then requires that factor structures be the same in different populations (Tisak and Meredith, 1990).

While the structures indicated above represent in some sense requirements for meaningful comparisons of differences in factors over some kind of repeated measurement (either occasion or population), other structures among the covariances of the manifest variables may be of interest. For example, if several factors are considered at each occasion, a structure might indicate both the pattern of factors and uniquenesses. It may be that characteristics of the uniquenesses across occasions are of interest. Longitudinal factor analytic strategies can test these suppositions. When modeling covariances, one can test statements regarding variances of both the factors and uniquenesses.

Longitudinal factor models are often estimated using confirmatory factor algorithms (e.g. LISREL, COSAN, EQS). Consider the following matrix of covariances:

		v_1o_1	v_2o_1	v_3o_1	v_1o_2	v_2o_2	v_3o_2
	v_1o_1	S11^2					
OCC 1	v_2o_1	C21 11 S21^2					
	v_3o_1	C31 11 C31 21 S31^2					

$$(1)$$

	v_1o_2	C12 11 C12 21 C12 31			S12^2		
OCC 2	v_2o_2	C22 11 C22 21 C22 31			C22 12 S22^2		
	v_3o_2	C32 11 C32 21 C32 31			C32 12 C32 22 S32^2		

where v_ko_t represents variable k at occasion t. A factor pattern can be used to represent the structure expected from the covariances. If the three variables measured are assumed to be indicators of the same latent construct, the following confirmatory pattern is expected:

	Factor I (occasion 1)	Factor II (occasion 2)
v_1o_1	p_1	0
v_2o_1	p_2	0
v_3o_1	p_3	0

$$(2)$$

$$
\begin{array}{ccc}
v_1o_2 & 0 & p_1 \\
v_2o_2 & 0 & p_2 \\
v_3o_2 & 0 & p_3
\end{array}
$$

The loadings are constrained equal across occasions. They reflect the relative size of the covariances within the occasion triangles in (1). These equality constraints require the pattern of covariances to be the same for both occasions.

Variables are required to load on the appropriate occasion factors (as shown above). As a result, cross-occasion covariances are modeled in the covariance matrix of the factors. The pattern of covariances in the cross-occasion rectangle determines the autocorrelation of the latent construct.

One major difference between longitudinal factor models and cross-sectional models appears in the pattern of residual covariances. In cross-sectional models the covariances among the uniquenesses are expected to be 0. The covariances between the pairs of repeatedly measured variables are, however, not independent. They will be higher than one would expect for two variables loading on different factors. The factors can be adjusted for these high values by allowing correlated residuals between each of these pairs. The covariance matrix of uniquenesses would be

	v_1o_1	v_2o_1	v_3o_1	v_1o_2	v_2o_2	v_3o_2
v_1o_1	u_{11}^2					
v_2o_1		u_{21}^2				
v_3o_1			u_{31}^2			
v_1o_2	$u_{1\,12}$			u_{21}^2		
v_2o_2		$u_{2\,12}$			u_{22}^2	
v_3o_2			$u_{3\,12}$			u_{32}^2

$$(3)$$

The factor pattern matrix, factor covariance matrix, the covariance matrix of uniquenesses when estimated yield the longitudinal autocorrelated factor model. This represents one of many models that may be of interest.

In this chapter we will fit a set of models suggested by Tisak and Meredith (1990) and McArdle and Aber (1990). These models will represent increasing degrees of constraint on the covariances of a set of indicators of a single latent construct measured repeatedly.

We will start with a baseline model that posits the simplest possible structure at three points in time, namely, one factor for each occasion with no restrictions on the pattern of loadings. We will then require that stationarity (invariance of structure across occasion) hold. That will be followed by an attempt to structure the unique variances equal across occasions. Next we add the restriction that the common factors have equal variances. We will last test invariance across groups in a two-group model.

2.   Outline of the Computational Procedures

The different longitudinal factor models will be estimated using LISREL VI. LISREL VI (now in release VII) is probably the most widely used SEM program available. A number of characteristics make it among the most flexible programs available. For estimating factor models, LISREL includes two equations. The x-side model is designed to handle pure factor analysis. The y-side model includes both a factor model and a matrix to look at all possible regressions among the latent variables. Two separate factor models can be estimated simultaneously (x-side and y-side). Relationships among latent variables can be estimated in yet another structural regression matrix. These multiple locations for estimating the same SEM contribute to LISREL's flexibility while very often adding to the confusion for the novice.

LISREL, along with many other SEM modeling programs, is currently available in both PC and mainframe versions.

3.   An Annotated Data Example

The data to be analyzed are three subtests from the Primary Mental Abilities (Thurstone and Thurstone, 1962): number facility (NUMF); letter series (LETS); and number series (NUMS). These data, described in Nesselroade and Baltes (1974), are taken from Tisak and Meredith (1990).

Step 1:   Preparation of the Data

A number of input data options are available in LISREL. The data can be entered as raw data or in the form of a symmetric matrix of association coefficients. If a correlation matrix is chosen as the input matrix, vectors of means and standard deviations can also be included. For this problem, we input a covariance matrix (mean levels of variables will not be tested in this analysis). Using a correlation matrix as input can result in a loss of important information. Since the correlation matrix is equivalent to a covariance matrix of z-scores, information about the change in variance from occasion to occasion is lost. In part because the scaling units of a covariance solution make the results difficult to interpret, a number of rescaling strategies that maintain the relative differences of observed score variables have been sug-

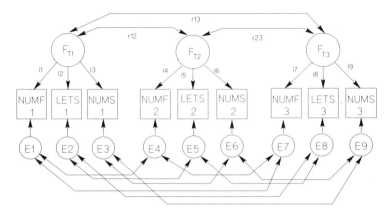

FIGURE 3.1.   Longitudinal factor analysis model.

gested. Tisak and Meredith (1990) suggest a number of ways to get an interpretable solution. Each generates a rescaled covariance matrix that looks something like a correlation matrix but maintains information about the relative variances. Among their suggestions are the following:

1. Select a single occasion from a single population and standardize all occasions for all populations to those values so that, if one were standardizing all occasions, **t**, and all populations, **p**, to those of the first occasion and first population, the rescaled coefficients would be

$$s_{xy\ tp}\ (\text{rescaled}) = s_{xy\ tp}/(s_{xx\ 11} * s_{yy\ 11}). \tag{4}$$

This scheme will yield a correlation matrix for the first occasion and population and something close to but not a correlation matrix for other occasions and populations (Tisak and Meredith, 1990).

2. Rescale using the reciprocal square of the average variance (Jöreskog, 1973) by first calculating the average variance for each measure (across occasion and population) and use those in place of the denomiator in equation 1. This scheme yields a scaled covariance matrix that can be partitioned into three occasion matrices and three cross-time matrices. The average of the three within occasion matrices would yield a correlation matrix in the rescaling. The latter rescaling was used by Tisak and Meredith (1990) to generate the input matrix for this data example.

Step 2:   Program Input (Test of Stationarity)

The model to be estimated is pictured in Figure 1. This represents a one-group longitudinal factor analysis model with no restrictions on the pattern of loadings. The input follows:

```
USERPROC NAME=LISREL
TITLE LONGITUDINAL FACTOR ANALYSIS PMA HORN & CATTELL 121 FEMALES
DA NG=1 NI=9 NO=121 MA=CM
LABEL
*
'NUMF1' 'LETS1' 'NUMS1' 'NUMF2' 'LETS2' 'NUMS2'
'NUMF3' 'LETS3' 'NUMS3'
CM SY
*
.789
.520 1.057
.511 .568 .892
.704 .502 .556 1.037
.454 .695 .469 .512 .849
.403 .484 .466 .526 .458 .824
.667 .579 .609 .847 .571 .527 1.028
.459 .762 .460 .528 .734 .488 .577 1.040
.462 .592 .534 .580 .568 .588 .631 .641 1.015
SELECT
1,2,3,4,5,6,7,8,9/
MO NX=9  NK=3 PH=SY,FI TD=SY,FI LX=FU,FI
PA LX
*
1 0 0
0 0 0
1 0 0
0 1 0
0 0 0
0 1 0
0 0 1
0 0 0
0 0 1
PA PH
*
1
1 1
1 1 1
PA TD
*
1
0 1
0 0 1
1 0 0 1
0 1 0 0 1
0 0 1 0 0 1
1 0 0 1 0 0 1
0 1 0 0 1 0 0 1
0 0 1 0 0 1 0 0 1
ST 1.0 LX 2 1 LX 5 2 LX 8 3
OU MI TV RS SS TO
END USER
```

The input data matrix is a scaled covariance matrix as described above. The pattern matrix describing the factors (**PA LX**) has one value (**LETS**) fixed at 1 for each factor. This scales the factor and gives the other factor loadings values relative to 1. The covariance matrix of the factor scores (**PHI**) is freely estimated (including the factor score variances). This means that the **PHI** matrix would have to be rescaled to yield the factor intercorrelations. The covariance matrix of residuals (**THETA DELTA**) estimates the uniquenesses in the diagonals. The off-diagonal estimates in **THETA DELTA** represent the expectation that uniquenesses should be autocorrelated; that is, the unique component of any variable should be correlated with a repeated administration of that same variable.

Step 3: Program Output

The input program generated the parameter estimates and fit indices:

```
TITLE LONGITUDINAL FACTOR ANALYSIS PMA HORN & CATTELL 121 FEMALES

LISREL ESTIMATES (MAXIMUM LIKELIHOOD)
```

LAMBDA X

	KSI 1	KSI 2	KSI 3
NUMF1	0.851	0.000	0.000
LETS1	1.000	0.000	0.000
NUMS1	0.974	0.000	0.000
NUMF2	0.000	1.082	0.000
LETS2	0.000	1.000	0.000
NUMS2	0.000	0.955	0.000
NUMF3	0.000	0.000	1.115
LETS3	0.000	0.000	1.000
NUMS3	0.000	0.000	1.101

PHI

	KSI 1	KSI 2	KSI 3
KSI 1	0.604		
KSI 2	0.501	0.485	
KSI 3	0.524	0.504	0.533

THETA DELTA

	NUMF1	LETS1	NUMS1	NUMF2	LETS2	NUMS2
NUMF1	0.349					
LETS1	0.000	0.455				
NUMS1	0.000	0.000	0.319			
NUMF2	0.241	0.000	0.000	0.466		
LETS2	0.000	0.196	0.000	0.000	0.367	
NUMS2	0.000	0.000	0.001	0.000	0.000	0.381
NUMF3	0.163	0.000	0.000	0.237	0.000	0.000
LETS3	0.000	0.246	0.000	0.000	0.233	0.000
NUMS3	0.000	0.000	-0.024	0.000	0.000	0.059

THETA DELTA

	NUMF3	LETS3	NUMS3
NUMF3	0.364		
LETS3	0.000	0.510	
NUMS3	0.000	0.000	0.370

SQUARED MULTIPLE CORRELATIONS FOR X - VARIABLES

NUMF1	LETS1	NUMS1	NUMF2	LETS2	NUMS2
0.557	0.570	0.642	0.550	0.567	0.537

NUMF3	LETS3	NUMS3
0.646	0.510	0.636

TOTAL COEFFICIENT OF DETERMINATION FOR X - VARIABLES IS 0.934

```
        MEASURES OF GOODNESS OF FIT FOR THE WHOLE MODEL :
CHI-SQUARE WITH   15 DEGREES OF FREEDOM IS     10.17 (PROB. LEVEL = 0.809)
              GOODNESS OF FIT INDEX IS 0.981
          ADJUSTED GOODNESS OF FIT INDEX IS 0.972
          ROOT MEAN SQUARE RESIDUAL IS      0.018
```

FITTED MOMENTS AND RESIDUALS

FITTED MOMENTS

	NUMF1	LETS1	NUMS1	NUMF2	LETS2	NUMS2
NUMF1	0.787					
LETS1	0.514	1.059				
NUMS1	0.500	0.588	0.891			
NUMF2	0.702	0.542	0.527	1.034		
LETS2	0.426	0.696	0.487	0.525	0.852	
NUMS2	0.407	0.478	0.466	0.501	0.463	0.824
NUMF3	0.660	0.584	0.569	0.845	0.562	0.537
LETS3	0.446	0.770	0.510	0.545	0.737	0.482
NUMS3	0.491	0.577	0.537	0.600	0.555	0.590

FITTED MOMENTS

	NUMF3	LETS3	NUMS3
NUMF3	1.027		
LETS3	0.594	1.043	
NUMS3	0.654	0.587	1.016

FITTED RESIDUALS

	NUMF1	LETS1	NUMS1	NUMF2	LETS2	NUMS2
NUMF1	0.002					
LETS1	0.006	-0.002				
NUMS1	0.011	-0.020	0.001			
NUMF2	0.002	-0.040	0.029	0.003		
LETS2	0.028	-0.001	-0.018	-0.013	-0.003	
NUMS2	-0.004	0.006	0.000	0.025	-0.005	0.000
NUMF3	0.007	-0.005	0.039	0.002	0.009	-0.010
LETS3	0.013	-0.008	-0.050	-0.017	-0.003	0.006
NUMS3	-0.029	0.015	-0.003	-0.020	0.013	-0.002

FITTED RESIDUALS

	NUMF3	LETS3	NUMS3
NUMF3	0.001		
LETS3	-0.017	-0.003	
NUMS3	-0.023	0.054	-0.001

NORMALIZED RESIDUALS

	NUMF1	LETS1	NUMS1	NUMF2	LETS2	NUMS2
NUMF1	0.022					
LETS1	0.065	-0.012				
NUMS1	0.121	-0.191	0.005			
NUMF2	0.020	-0.368	0.288	0.022		
LETS2	0.331	-0.014	-0.202	-0.130	-0.031	
NUMS2	-0.049	0.060	-0.002	0.258	-0.062	-0.002
NUMF3	0.071	-0.047	0.387	0.019	0.088	-0.104
LETS3	0.143	-0.066	-0.502	-0.162	-0.027	0.066
NUMS3	-0.309	0.142	-0.030	-0.188	0.131	-0.016

NORMALIZED RESIDUALS

	NUMF3	LETS3	NUMS3
NUMF3	0.009		
LETS3	-0.160	-0.022	
NUMS3	-0.211	0.501	-0.004

The chi-square goodness of fit tests whether the sample covariance matrix could have been drawn from the population represented by the population covariance matrix estimated by the model. The input matrix has $\mathbf{K}=\mathbf{k(k+1)}$ /2 degrees of freedom where **k** is the dimension of the covariance matrix. The test is distributed over $\mathbf{K-P}$ degrees of freedom where **P** is the number of unique parameters estimated by the model (for the first model $\mathbf{K}=45$ and $\mathbf{P}=30$ for $\mathbf{df}=15$). A non-significant *chi*-square suggests that the estimated population matrix and the sample matrix are not different.

The goodness-of-fit index has values ranging from 0 to 1 where the larger value represents the better fit. This index, along with the adjusted goodness-of-fit and other indices of fit, is discussed by Bentler and Bonnet (1980) and Hoelter (1983). McArdle and Aber (1990) discussed the use of fit indices to assess the relative fit of models.

The root mean square residual can be interpreted as an indication of the average size of the difference between a sample covariance coefficient and the corresponding estimated population coefficient.

The values of the estimated population matrix (**SIGMA**) are listed. These are computed by constraining parameters fixed by the model and allowing unrestricted parameters to be estimated in a way that maximizes the agreement with the observed matrix. These are followed by the fitted residuals. These values are normalized so that the relative size of residuals can be considered on a common scale. Normalized residuals can be interpreted as z-scores. A 95% confidence interval could be build around the residual using $z \pm 1.96$. For a single residual, a value of the normalized residual larger than 1.96 represents a residual that could be considered different than zero. Some caution is suggested because of the large number of residuals that may sit in the matrix. Some are expected to be non-zero by chance. The normalized residuals will point the finger at possible relationships the model is not explaining. These residuals tend to be somewhat less affected by sample size than the modification indices, which estimate how much the chi-square goodness-of-fit will change if a restriction on a parameter is relaxed.

The fit of the model showed that the structure anticipated by the model agreed with the data. This was to be expected, as the model is essentially unrestricted.

The next question addressed considered whether stationarity was a viable hypotheses. To test stationarity the following constraints were added to the model:

```
EQ LX 1 1 LX 4 2 LX 7 3
EQ LX 3 1 LX 6 2 LX 9 3
```

These constraints yielded the following factor solution and fit indices:

TITLE LONGITUDINAL FACTOR ANALYSIS PMA HORN & CATTELL 121 FEMALES

LISREL ESTIMATES (MAXIMUM LIKELIHOOD)

LAMBDA X

	KSI 1	KSI 2	KSI 3
NUMF1	0.991	0.000	0.000
LETS1	1.000	0.000	0.000
NUMS1	1.006	0.000	0.000
NUMF2	0.000	0.991	0.000
LETS2	0.000	1.000	0.000
NUMS2	0.000	1.006	0.000
NUMF3	0.000	0.000	0.991
LETS3	0.000	0.000	1.000
NUMS3	0.000	0.000	1.006

PHI

	KSI 1	KSI 2	KSI 3
KSI 1	0.514		
KSI 2	0.472	0.506	
KSI 3	0.515	0.540	0.605

THETA DELTA

	NUMF1	LETS1	NUMS1	NUMF2	LETS2	NUMS2
NUMF1	0.331					
LETS1	0.000	0.469				
NUMS1	0.000	0.000	0.344			
NUMF2	0.233	0.000	0.000	0.480		
LETS2	0.000	0.191	0.000	0.000	0.356	
NUMS2	0.000	0.000	0.006	0.000	0.000	0.374
NUMF3	0.155	0.000	0.000	0.251	0.000	0.000
LETS3	0.000	0.235	0.000	0.000	0.218	0.000
NUMS3	0.000	0.000	-0.008	0.000	0.000	0.057

THETA DELTA

	NUMF3	LETS3	NUMS3
NUMF3	0.379		
LETS3	0.000	0.490	
NUMS3	0.000	0.000	0.378

SQUARED MULTIPLE CORRELATIONS FOR X - VARIABLES

NUMF1	LETS1	NUMS1	NUMF2	LETS2	NUMS2
0.581	0.556	0.615	0.537	0.581	0.546

NUMF3	LETS3	NUMS3
0.631	0.529	0.627

TOTAL COEFFICIENT OF DETERMINATION FOR X - VARIABLES IS 0.925

MEASURES OF GOODNESS OF FIT FOR THE WHOLE MODEL :
CHI-SQUARE WITH 19 DEGREES OF FREEDOM IS   16.91 (PROB. LEVEL = 0.596)
GOODNESS OF FIT INDEX IS 0.970
ADJUSTED GOODNESS OF FIT INDEX IS 0.948
ROOT MEAN SQUARE RESIDUAL IS   0.044

Once again the fit indicated a tenable hypothesis.

To further restrict the model we next tested whether unique variances for a particular variable were the same across occasion by adding the following restriction:

```
EQ TD 1 1 TD 4 4 TD 7 7
EQ TD 2 2 TD 5 5 TD 8 8
EQ TD 3 3 TD 6 6 TD 9 9
```

The model and fit were

```
TITLE LONGITUDINAL FACTOR ANALYSIS PMA HORN & CATTELL 121 FEMALES

LISREL ESTIMATES (MAXIMUM LIKELIHOOD)
```

LAMBDA X

		KSI 1	KSI 2	KSI 3
NUMF1		1.003	0.000	0.000
LETS1		1.000	0.000	0.000
NUMS1		1.024	0.000	0.000
NUMF2		0.000	1.003	0.000
LETS2		0.000	1.000	0.000
NUMS2		0.000	1.024	0.000
NUMF3		0.000	0.000	1.003
LETS3		0.000	0.000	1.000
NUMS3		0.000	0.000	1.024

PHI

	KSI 1	KSI 2	KSI 3
KSI 1	0.502		
KSI 2	0.469	0.510	
KSI 3	0.508	0.544	0.600

THETA DELTA

	NUMF1	LETS1	NUMS1	NUMF2	LETS2	NUMS2
NUMF1	0.399					
LETS1	0.000	0.447				
NUMS1	0.000	0.000	0.349			
NUMF2	0.231	0.000	0.000	0.399		
LETS2	0.000	0.216	0.000	0.000	0.447	
NUMS2	0.000	0.000	-0.008	0.000	0.000	0.349
NUMF3	0.183	0.000	0.000	0.229	0.000	0.000
LETS3	0.000	0.216	0.000	0.000	0.242	0.000
NUMS3	0.000	0.000	-0.022	0.000	0.000	0.038

THETA DELTA

	NUMF3	LETS3	NUMS3
NUMF3	0.399		
LETS3	0.000	0.447	
NUMS3	0.000	0.000	0.349

SQUARED MULTIPLE CORRELATIONS FOR X - VARIABLES

NUMF1	LETS1	NUMS1	NUMF2	LETS2	NUMS2
0.494	0.578	0.609	0.615	0.474	0.577

NUMF3	LETS3	NUMS3
0.612	0.571	0.656

```
TOTAL COEFFICIENT OF DETERMINATION FOR X - VARIABLES IS  0.926


                  MEASURES OF GOODNESS OF FIT FOR THE WHOLE MODEL :
  CHI-SQUARE WITH  25 DEGREES OF FREEDOM IS      25.07 (PROB. LEVEL = 0.459)
                     GOODNESS OF FIT INDEX IS 0.956
                  ADJUSTED GOODNESS OF FIT INDEX IS 0.902
                   ROOT MEAN SQUARE RESIDUAL IS      0.053
```

At this point we could have continued to test more and more restricted models. For example, we could next require that all unique variances be equal. We could similarly begin to place restrictions on the factor covariances. For example, we could test a model of no association among the factor scores, or we could test for a simplex structure represented by equal adjacent covariances ($r_{occ1.occ2}$, $r_{occ2.occ3}$). However, at this point we consider the question of invariance across populations.

Step 4:   Program and Output for Invariance Across Groups

To test invariance across groups we ran a two-group model constraining the factor pattern to be equivalent across groups. To test this model we used the following input:

```
USERPROC NAME=LISREL
TITLE LONGITUDINAL FACTOR ANALYSIS PMA HORN & CATTELL 121 FEMALES
DA NG=2 NI=9 NO=121 MA=CM
LABEL
*
'NUMF1' 'LETS1' 'NUMS1' 'NUMF2' 'LETS2' 'NUMS2'
'NUMF3' 'LETS3' 'NUMS3'
CM SY
*
.782
.520 1.057
.511 .968 .392
.704 .502 .556 1.037
.454 .695 .169 .512 .849
.403 .484 .166 .526 .458 .824
.667 .579 .608 .847 .571 .527 1.028
.459 .762 .160 .523 .734 .488 .577 1.040
.462 .592 .534 .580 .568 .588 .631 .641 1.015
SELECT
1,2,3,4,5,6,7,8,9/
MO NX=9  NK=3 PH=SY,FI TD=SY,FI LX=FU,FI
PA LX
*
1 0 0
0 0 0
1 0 0
0 1 0
0 0 0
0 1 0
0 0 1
0 0 0
0 0 1
PA PH
*
1
1 1
1 1 1
```

```
EQ LX 1 1 LX 4 2 LX 7 3
PA TD
*
1
0 1
0 0 1
1 0 0 1
0 1 0 0 1
0 0 1 0 0 1
1 0 0 1 0 0 1
0 1 0 0 1 0 0 1
0 0 1 0 0 1 0 0 1
EQ LX 3 1 LX 6 2 LX 9 3
ST 1.0 LX 2 1 LX 5 2 LX 8 3
OU MI
TITLE PMA 93 MALES
DA NO=93
LABEL
*
'NUMF1' 'LETS1' 'NUMS1' 'NUMF2' 'LETS2' 'NUMS2'
'NUMF3' 'LETS3' 'NUMS3'
CM SY
*
.745
.414 .857
.453 .417 .901
.686 .562 .527 1.056
.484 .658 .516 .747 1.194
.540 .512 .592 .684 .697 1.110
.741 .715 .617 .905 .807 .821 1.389
.482 .453 .495 .596 .764 .646 .776 1.020
.631 .616 .629 .824 .841 .900 1.026 .823 1.339
SELECT
1,2,3,4,5,6,7,8,9
MO LX=IN PH=PS TD=PS
OU MI TV RS SS TO
END USER
```

The equivalence statements (EQ) applied the constraints across groups. The parameter estimates and fit of the model were

```
TITLE LONGITUDINAL FACTOR ANALYSIS PMA HORN & CATTELL 121 FEMALES

LISREL ESTIMATES (MAXIMUM LIKELIHOOD)
```

	LAMBDA X		
	KSI 1	KSI 2	KSI 3
NUMF1	1.082	0.000	0.000
LETS1	1.000	0.000	0.000
NUMS1	1.085	0.000	0.000
NUMF2	0.000	1.082	0.000
LETS2	0.000	1.000	0.000
NUMS2	0.000	1.085	0.000
NUMF3	0.000	0.000	1.082
LETS3	0.000	0.000	1.000
NUMS3	0.000	0.000	1.085

	PHI		
	KSI 1	KSI 2	KSI 3
KSI 1	0.448		
KSI 2	0.418	0.454	
KSI 3	0.452	0.489	0.536

```
        THETA DELTA
```

	NUMF1	LETS1	NUMS1	NUMF2	LETS2	NUMS2
NUMF1	0.330					
LETS1	0.000	0.492				
NUMS1	0.000	0.000	0.336			
NUMF2	0.231	0.000	0.000	0.471		
LETS2	0.000	0.205	0.000	0.000	0.370	
NUMS2	0.000	0.000	-0.002	0.000	0.000	0.367
NUMF3	0.154	0.000	0.000	0.243	0.000	0.000
LETS3	0.000	0.250	0.000	0.000	0.232	0.000
NUMS3	0.000	0.000	-0.015	0.000	0.000	0.051

THETA DELTA

	NUMF3	LETS3	NUMS3
NUMF3	0.372		
LETS3	0.000	0.505	
NUMS3	0.000	0.000	0.373

SQUARED MULTIPLE CORRELATIONS FOR X - VARIABLES

NUMF1	LETS1	NUMS1	NUMF2	LETS2	NUMS2
0.557	0.426	0.627	0.554	0.690	0.669

NUMF3	LETS3	NUMS3
0.732	0.505	0.721

TOTAL COEFFICIENT OF DETERMINATION FOR X - VARIABLES IS   0.972

MEASURES OF GOODNESS OF FIT FOR THE WHOLE MODEL :
GOODNESS OF FIT INDEX IS 0.969
ROOT MEAN SQUARE RESIDUAL IS   0.049

TITLE PMA 93 MALES

LISREL ESTIMATES (MAXIMUM LIKELIHOOD)

LAMBDA X

	KSI 1	KSI 2	KSI 3
NUMF1	1.082	0.000	0.000
LETS1	1.000	0.000	0.000
NUMS1	1.085	0.000	0.000
NUMF2	0.000	1.082	0.000
LETS2	0.000	1.000	0.000
NUMS2	0.000	1.085	0.000
NUMF3	0.000	0.000	1.082
LETS3	0.000	0.000	1.000
NUMS3	0.000	0.000	1.085

PHI

	KSI 1	KSI 2	KSI 3
KSI 1	0.364		
KSI 2	0.456	0.632	
KSI 3	0.524	0.682	0.302

THETA DELTA

	NUMF1	LETS1	NUMS1	NUMF2	LETS2	NUMS2
NUMF1	0.326					
LETS1	0.000	0.430				
NUMS1	0.000	0.000	0.482			
NUMF2	0.169	0.000	0.000	0.345		
LETS2	0.000	0.128	0.000	0.000	0.474	
NUMS2	0.000	0.000	0.077	0.000	0.000	0.415
NUMF3	0.107	0.000	0.000	0.093	0.000	0.000
LETS3	0.000	-0.067	0.000	0.000	0.115	0.000
NUMS3	0.000	0.000	0.002	0.000	0.000	0.083

```
THETA DELTA

     NUMF3        LETS3       NUMS3
        ─────      ─────       ─────
NUMF3    0.373
LETS3    0.000      0.390
NUMS3    0.000      0.000       0.325

SQUARED MULTIPLE CORRELATIONS FOR X - VARIABLES

     NUMF1       LETS1       NUMS1      NUMF2      LETS2      NUMS2
     ─────       ─────       ─────      ─────      ─────      ─────
     0.563       0.499       0.465      0.673      0.603      0.626
     NUMF3       LETS3       NUMS3
     ─────       ─────       ─────
     0.731       0.618       0.757

TOTAL COEFFICIENT OF DETERMINATION FOR X - VARIABLES IS   0.941

          MEASURES OF GOODNESS OF FIT FOR THE WHOLE MODEL :
CHI-SQUARE WITH  40 DEGREES OF FREEDOM IS      35.18 (PROB. LEVEL = 0.687)
             GOODNESS OF FIT INDEX IS 0.960
          ROOT MEAN SQUARE RESIDUAL IS      0.063
```

showing that the hypotheses of stationarity and invariance across groups was reasonable given the observed data.

4. Discussion

The test of the models suggested that the factor structure satisfied the criteria of stationarity and invariance across groups. This means it is reasonable to consider that the latent construct captured by the indicators has the same structure at different occasions and in different (at least two) groups. At this point, it becomes reasonable to talk about change in the construct. The powerful methods of structural equation modeling (SEM) can now be used to explore questions about change among the latent variables (McArdle and Aber, 1990).

When longitudinal factor analysis is used to establish a best measure of the latent constructs, it is establishing a plausible measurement model for the repeatedly measured observed variables. The covariances generated by the model then represent a matrix of associations that can be "passed on" to further analysis either by moving to a more complex SEM or by analyzing the matrix of factor covariances. Structures among the factor covariances could now be tested. As an example, these covariances could be considered grist for an autoregressive model. The test of the simplex structure in the factor covariances becomes an interpretable test. If the analysis was done using an augmented matrix of cross-products, tests regarding the level of latent variables could be considered (McArdle and Aber, 1990). Once factor invariance has been established, the many change analysis strategies become more easily interpretable.

Longitudinal factor analytic strategies can be used to test hypotheses about patterns of change. Consider three variables measured on three occasions. Suppose one was interested in showing that the pattern or shape of change was the same for both variables. One could turn each set of three variables into a new set representing an overall mean, a linear trend, and a quadratic trend (Belsky and Rovine, 1990). It would be possible to include the level and trend scores as new variables in a factor analysis. (McArdle and Nesselroade, 1988). If the levels, the linear trends, and the quadratic trends tend to load respectively on separate factors, they would indicate that the three variables are changing in somewhat the same fashion.

Longitudinal factor analytic strategies represent powerful methods both for considering measurement characteristics of latent variables and for delineating patterns of change among observed variables.

Chapter 4

Latent Variable Structural Equations Models

1. Summary of Method

Structural equation models (SEMs) can be used to estimate a variety of models describing change in a longitudinal data set. The models most often considered under the term SEM are confirmatory factor analysis models (Tisak and Meredith, 1990: McArdle and Aber, 1990), structural regression models with single indicator observed variables, and structural regression models with both latent and observed variable in a variety of combinations (McArdle and Aber, 1990; Goldberger and Duncan, 1973). These models have begun to appear in many different disciplines (see Hayduk, 1987; Bollen, 1989; Joreskog and Wold, 1982).

A number of basic, qualitatively different models are available to the researcher fortunate enough to have multiple indicators of the same construct measured on multiple occasions. All are in some sense structural models (Goldberger and Duncan, 1973) in that they try to explain the pattern of relationships in a covariance structure based on a set of hypothesized relationships among variables, both observed and latent. They are path models in the sense that a general method of diagramming these models based on well-known rules (Joreskog, 1973; McArdle and McDonald, 1984) can lead to a set of estimation equations for the parameters of the model. The models

55

differ in the kinds of statements they make about change. In this sense the models are different, not necessarily competing.

We will deal here with an example of each of three general model types described by McArdle and Aber (1990): autoregressive, difference component, and latent growth-curve models.

The selection of an autoregressive model often depends on the researchers interest in stability patterns in the data. They are particularly useful when one wants to make regression-type predictions based on prior values of the same or other variables. Difference component models posit latent variables that determine the change in the variability of a process across two or more occasions. Latent growth curve models test whether an underlying process determines the shape of growth or a trend in a set of repeated measures.

As examples of the three basic model types, we will compute the age-based Markov model (autoregressive), a moving average difference model (difference component), and a linear growth model (latent growth curve).

The simplest autoregressive models estimate scores on a variable based on prior values of the same variable. A model in which scores for a variable at time, t_k, is predicted only by the score for that variable at time, t_{k-1}, is termed a first order autoregressive or Markov simplex model (McArdle and Aber, 1990; Hertzog and Nesselroade, 1987). The model we compute is depicted in Figure 1. This figure has been drawn using RAM notation (McArdle and Horn, 1988) and represents a model in which a variable (WISC) has been measured at four different time points (represented as squares) that are unequally spaced (6, 7, 9, and 11 years). The model assumes, however, that the researchers wished they had measured each participant yearly. The unobserved variables in the autoregressive chain (represented as circles) are those missing time points. Each observed variable, y_k, is in part determined by an unobserved disturbance term, e_k, which has a variance, vU_k. Each unobserved variable, D_k, also has a variance, vD_k. In this model the path coefficient representing prediction from one time to the next is equal between adjacent measures and is given the value **p**. The variances of the the measures across occasion are to be constrained to be equal.

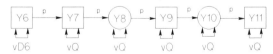

FIGURE 4.1.   Age-based Markov process.

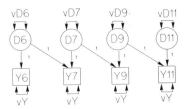

FIGURE 4.2.   Moving average difference model.

The difference component model selected is the moving average difference model (Figure 2). This model posits four unobserved components, D_k, each of which models the change between two time adjacent scores on a particular variable. The difference component is created by running two paths constrained to be equal and to a constant from the latent variable to the pair of adjacent observed variable. The variance of each component estimates the degree to which it describes the developmental process.

The linear latent growth curve model is depicted in Figure 3. The model estimates a curve for the four occasions of score values that have been collected. The shape of the curve, S, is either estimated or prescribed by the basis coeffients, b_k. The overall level of the curve can be indicated by using a latent variable, L, with a path coeffient value of 1 into each of the observed variables. The constant in the model is used to test the means of L and S. The standard deviation of these variables is represented as a path from external unobserved variables.

The strategies involved in estimating each of these models require 1) an adequate description of the problem (represented diagrammatically for these models); 2) the determination of the set of equations necessary to estimate the parameters of interest; 3) the determination of the most appropriate input for the problem; and 4) the selection of an algorithm to estimate the model.

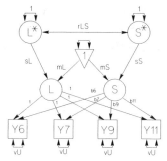

FIGURE 4.3.   Latent growth-curve model.

2. Outline of the Computational Procedures

Computation of SEMs requires an algorithm that can analyze symmetric matrices representing associative structures in data. These may be correlation, covariance, or moment matrices. As there are so many different structures that may be of interest, programs have been specifically designed to estimate SEMs. Estimations performed by the SEM programs can be done in other programming environments. It is relatively easy, for example, to estimate a single indicator path model using any regression procedure. It is much more difficult to get an exploratory factor analysis procedure to do a confirmatory factor analysis. As a result, SEM programs represent a convenience for users. Any SEM can be estimated by a good matrix algebra program.

There are currently a number of excellent programs for estimating SEMs: COSAN (Fraser and McDonald, 1988); EQS (Bentler, 1985); AMOS (Arbuckle, 1989); and LISREL VII (Joreskog and Sorbom, 1989). All of these programs are currently in microcomputer versions. We selected LISREL to estimate the models, in part because of its wide availability and popularity. Some of these models, however, may be more easily handled using other programs (McArdle and Aber, 1990, technical appendix).

3. Annotated Data Examples for Each Model

AGE-BASED MARKOV MODEL

Step 1: Preparation of the Data

The data selected for this problem are the WISC data reported in McArdle and Epstein (1987; McArdle and Aber, 1990). Measures of the WISC taken at 6, 7, 9, and 11 years of age are used in this model. The form of the data input for the autoregressive problem is a correlation matrix along with vectors of means and standard deviations. The LISREL input statements are

```
USERPROC NAME=LISREL
TITLE AGE-BASED MARKOV PROCESS
DA NG=1 NI=6 NOBS=204 MA=CM
LABELS
*
'WISC_6' 'WISC_7' 'WISC_9' 'WISC_11' 'MOTHERED' 'CONSTANT'
ME
    18.034 25.819 35.255 46.593 10.311 1.000
SD
    6.734 7.319 7.796 10.386 2.700 0.000
KM SY
```

```
*
1.000
 .809 1.000
 .806  .850 1.000
 .765  .831  .867 1.000
 .520  .531  .448  .458 1.000
 .000  .000  .000  .000  .000 1.000
SE
 1 2 3 4/
```

Step 2: The Program (Adapted from McArdle and Aber, 1990)

The model in Figure 1 has four **Y** variables and 12 **ETA** variables. This setup may be confusing to those unfamiliar with RAM notation, but the logic of the model is straightforward. Each variable, whether it is latent or observed, gets a place in the **BETA** matrix. **BETA** values model all relationships among variables regardless of whether they are observed or latent. So for this model, the **BETA** matrix is the home of four observed variables, **WISC__6**, **WISC__7**, **WISC__9**, and **WISC__11**, two unobserved variables representing the missing observations, **WISC__8** and **WISC__10**, four unobserved disturbance variables contributed to the variation in the observed variable—**ERR__6**, **ERR__7**, **ERR__9**, and **ERR__11**—and two external variables contributing to the variance of the unobserved measurement point, **ERR__8**, and **ERR__11**.

Parameter estimates representing path relationships will occur in the **BETA** matrix, and variance estimates will appear in the **PSI** matrix. The **LAMBDA** matrix will be used to determine which of the **ETA**s are observed. The LISREL program for this model follows.

```
MO NY=4 NE=12 BE=FI,FU PS=FI,SY LY=FI TE=ZE
LE
  '1.WISC_6' '2.WISC_7' '3.WISC_8' '4.WISC_9' '5.WISC10' '6.WISC11'
  '7.ERR.6' '8.ERR_7' '9.ERR_8' '10.ERR_9' '11.ERR10' '12.ERR11'
MA LY
*
1 0 0 0 0 0 0 0 0 0 0 0
0 1 0 0 0 0 0 0 0 0 0 0
0 0 0 1 0 0 0 0 0 0 0 0
0 0 0 0 0 1 0 0 0 0 0 0
MA BE
*
0 0 0 0 0 0 1 0 0 0 0 0
1 0 0 0 0 0 0 1 0 0 0 0
0 1 0 0 0 0 0 0 1 0 0 0
0 0 1 0 0 0 0 0 0 1 0 0
0 0 0 1 0 0 0 0 0 0 1 0
0 0 0 0 1 0 0 0 0 0 0 1
0 0 0 0 0 0 0 0 0 0 0 0
0 0 0 0 0 0 0 0 0 0 0 0
0 0 0 0 0 0 0 0 0 0 0 0
0 0 0 0 0 0 0 0 0 0 0 0
0 0 0 0 0 0 0 0 0 0 0 0
0 0 0 0 0 0 0 0 0 0 0 0
MA PS
*
0
0 0
0 0 0
```

```
0 0 0 0
0 0 0 0 0
0 0 0 0 0 0
0 0 0 0 0 0 1
0 0 0 0 0 0 0 1
0 0 0 0 0 0 0 0 1
0 0 0 0 0 0 0 0 0 1
0 0 0 0 0 0 0 0 0 0 1
0 0 0 0 0 0 0 0 0 0 0 1
FR BE 2 1 BE 3 2 BE 4 3 BE 5 4 BE 6 5
FR BE 2 8 BE 3 9 BE 4 10 BE 5 11 BE 6 12
FR PS 1 1
EQ BE 2 1 BE 3 2 BE 4 3 BE 5 4 BE 6 5
EQ BE 2 8 BE 3 9 BE 4 10 BE 5 11 BE 6 12
OU PT SE TV NS MI SS RS ND=3 TO
END USER
```

This code may strike LISREL users unfamiliar with RAM notation as unusual. A good way to think about the model is to remember that all variables, both observed and unobserved, occupy a place in the **BETA** matrix. As a result, all arrows are modeled in that matrix. This includes the formation of factors, the contribution of residual terms, and causes of an unobserved variable external to the model. Residual variances estimates can appear in either the **PSI** matrix as variance estimates or in the **BETA** matrix as standard deviations represented by the path from a disturbance to an observed variable. In this example, standard deviations for the observed variables are located in **BETA 1 7 BETA 2 8 BETA 4 10** and **BETA 6 12**.

This model is unusual in that unobserved marker variables are included to "save the place" for uncollected occasions. This allows one to specify equal path coefficients for the now equally spaced intervals of measurement. Constraints (e.g., path coefficients before and after the unobserved occasion are equal) must be applied to parameter estimates to allow the model to be identified. In this model those constraints are applied to the path coefficients and the variances of both the collected and uncollected WISC scores.

Step 3: Output for Age-based Markov Model

For this first model, we include the portions of the output detailing the model along with the LISREL estimates and some of the accompanying output.

As a check of the model, the covariance matrix, parameters to be estimated, and the starting values specified by the MA LY, MA BE, and MA PS statements follow:

```
TITLE AGE-BASED MARKOV PROCESS

       COVARIANCE MATRIX TO BE ANALYZED

          WISC_6    WISC_7    WISC_9    WISC_11
         -------   -------   -------   -------
WISC_6    45.347
WISC_7    39.872    53.568
WISC_9    42.314    48.500    60.778
WISC_11   53.504    63.169    70.200   107.869
```

DETERMINANT = 0.279619D+06

PARAMETER SPECIFICATIONS

LAMBDA Y

	1.WISC_6	2.WISC_7	3.WISC_8	4.WISC_9	5.WISC10	6.WISC11
WISC_6	0	0	0	0	0	0
WISC_7	0	0	0	0	0	0
WISC_9	0	0	0	0	0	0
WISC_11	0	0	0	0	0	0

LAMBDA Y

	7.ERR.6	8.ERR_7	9.ERR_8	10.ERR_9	11.ERR10	12.ERR11
WISC_6	0	0	0	0	0	0
WISC_7	0	0	0	0	0	0
WISC_9	0	0	0	0	0	0
WISC_11	0	0	0	0	0	0

BETA

	1.WISC_6	2.WISC_7	3.WISC_8	4.WISC_9	5.WISC10	6.WISC11
1.WISC_6	0	0	0	0	0	0
2.WISC_7	1	0	0	0	0	0
3.WISC_8	0	1	0	0	0	0
4.WISC_9	0	0	1	0	0	0
5.WISC10	0	0	0	1	0	0
6.WISC11	0	0	0	0	1	0
7.ERR.6	0	0	0	0	0	0
8.ERR_7	0	0	0	0	0	0
9.ERR_8	0	0	0	0	0	0
10.ERR_9	0	0	0	0	0	0
11.ERR10	0	0	0	0	0	0
12.ERR11	0	0	0	0	0	0

BETA

	7.ERR.6	8.ERR_7	9.ERR_8	10.ERR_9	11.ERR10	12.ERR11
1.WISC_6	0	0	0	0	0	0
2.WISC_7	0	2	0	0	0	0
3.WISC_8	0	0	2	0	0	0
4.WISC_9	0	0	0	2	0	0
5.WISC10	0	0	0	0	2	0
6.WISC11	0	0	0	0	0	2
7.ERR.6	0	0	0	0	0	0
8.ERR_7	0	0	0	0	0	0
9.ERR_8	0	0	0	0	0	0
10.ERR_9	0	0	0	0	0	0
11.ERR10	0	0	0	0	0	0
12.ERR11	0	0	0	0	0	0

PSI

	1.WISC_6	2.WISC_7	3.WISC_8	4.WISC_9	5.WISC10	6.WISC11
1.WISC_6	3					
2.WISC_7	0	0				
3.WISC_8	0	0	0			
4.WISC_9	0	0	0	0		
5.WISC10	0	0	0	0	0	
6.WISC11	0	0	0	0	0	0
7.ERR.6	0	0	0	0	0	0
8.ERR_7	0	0	0	0	0	0
9.ERR_8	0	0	0	0	0	0
10.ERR_9	0	0	0	0	0	0
11.ERR10	0	0	0	0	0	0
12.ERR11	0	0	0	0	0	0

PSI

	7.ERR.6	8.ERR_7	9.ERR_8	10.ERR_9	11.ERR10	12.ERR11
7.ERR.6	0					
8.ERR_7	0	0				
9.ERR_8	0	0	0			
10.ERR_9	0	0	0	0		
11.ERR10	0	0	0	0	0	
12.ERR11	0	0	0	0	0	0

STARTING VALUES

LAMBDA Y

	1.WISC_6	2.WISC_7	3.WISC_8	4.WISC_9	5.WISC10	6.WISC11
WISC_6	1.000	0.000	0.000	0.000	0.000	0.000
WISC_7	0.000	1.000	0.000	0.000	0.000	0.000
WISC_9	0.000	0.000	0.000	1.000	0.000	0.000
WISC_11	0.000	0.000	0.000	0.000	0.000	1.000

LAMBDA Y

	7.ERR.6	8.ERR_7	9.ERR_8	10.ERR_9	11.ERR10	12.ERR11
WISC_6	0.000	0.000	0.000	0.000	0.000	0.000
WISC_7	0.000	0.000	0.000	0.000	0.000	0.000
WISC_9	0.000	0.000	0.000	0.000	0.000	0.000
WISC_11	0.000	0.000	0.000	0.000	0.000	0.000

BETA

	1.WISC_6	2.WISC_7	3.WISC_8	4.WISC_9	5.WISC10	6.WISC11
1.WISC_6	0.000	0.000	0.000	0.000	0.000	0.000
2.WISC_7	1.000	0.000	0.000	0.000	0.000	0.000
3.WISC_8	0.000	1.000	0.000	0.000	0.000	0.000
4.WISC_9	0.000	0.000	1.000	0.000	0.000	0.000
5.WISC10	0.000	0.000	0.000	1.000	0.000	0.000
6.WISC11	0.000	0.000	0.000	0.000	1.000	0.000
7.ERR.6	0.000	0.000	0.000	0.000	0.000	0.000
8.ERR_7	0.000	0.000	0.000	0.000	0.000	0.000
9.ERR_8	0.000	0.000	0.000	0.000	0.000	0.000
10.ERR_9	0.000	0.000	0.000	0.000	0.000	0.000
11.ERR10	0.000	0.000	0.000	0.000	0.000	0.000
12.ERR11	0.000	0.000	0.000	0.000	0.000	0.000

BETA

	7.ERR.6	8.ERR_7	9.ERR_8	10.ERR_9	11.ERR10	12.ERR11
1.WISC_6	1.000	0.000	0.000	0.000	0.000	0.000
2.WISC_7	0.000	1.000	0.000	0.000	0.000	0.000
3.WISC_8	0.000	0.000	1.000	0.000	0.000	0.000
4.WISC_9	0.000	0.000	0.000	1.000	0.000	0.000
5.WISC10	0.000	0.000	0.000	0.000	1.000	0.000
6.WISC11	0.000	0.000	0.000	0.000	0.000	1.000
7.ERR.6	0.000	0.000	0.000	0.000	0.000	0.000
8.ERR_7	0.000	0.000	0.000	0.000	0.000	0.000
9.ERR_8	0.000	0.000	0.000	0.000	0.000	0.000
10.ERR_9	0.000	0.000	0.000	0.000	0.000	0.000
11.ERR10	0.000	0.000	0.000	0.000	0.000	0.000
12.ERR11	0.000	0.000	0.000	0.000	0.000	0.000

PSI

	1.WISC_6	2.WISC_7	3.WISC_8	4.WISC_9	5.WISC10	6.WISC11
1.WISC_6	0.000					
2.WISC_7	0.000	0.000				
3.WISC_8	0.000	0.000	0.000			
4.WISC_9	0.000	0.000	0.000	0.000		

5.WISC10	0.000	0.000	0.000	0.000	0.000	
6.WISC11	0.000	0.000	0.000	0.000	0.000	0.000
7.ERR.6	0.000	0.000	0.000	0.000	0.000	0.000
8.ERR_7	0.000	0.000	0.000	0.000	0.000	0.000
9.ERR_8	0.000	0.000	0.000	0.000	0.000	0.000
10.ERR_9	0.000	0.000	0.000	0.000	0.000	0.000
11.ERR10	0.000	0.000	0.000	0.000	0.000	0.000
12.ERR11	0.000	0.000	0.000	0.000	0.000	0.000

PSI

	7.ERR.6	8.ERR_7	9.ERR_8	10.ERR_9	11.ERR10	12.ERR11
7.ERR.6	1.000					
8.ERR_7	0.000	1.000				
9.ERR_8	0.000	0.000	1.000			
10.ERR_9	0.000	0.000	0.000	1.000		
11.ERR10	0.000	0.000	0.000	0.000	1.000	
12.ERR11	0.000	0.000	0.000	0.000	0.000	1.000

This model estimates only two unique parameters. The counts in the parameter definition matrices serve to check whether the input is estimating the model of concern. The starting values in this case are bad but adequate guesses of the final values of the parameters. (NOTE: LISREL checks each matrix that can function as a variance/covariance for positive-definiteness. The warning regarding the **PSI** matrix can be ignored since the model requires certain of the diagonal elements to be **0**).

The LISREL estimates representing the results of the model are

LISREL ESTIMATES (MAXIMUM LIKELIHOOD)

LAMBDA Y

	1.WISC_6	2.WISC_7	3.WISC_8	4.WISC_9	5.WISC10	6.WISC11
WISC_6	1.000	0.000	0.000	0.000	0.000	0.000
WISC_7	0.000	1.000	0.000	0.000	0.000	0.000
WISC_9	0.000	0.000	0.000	1.000	0.000	0.000
WISC_11	0.000	0.000	0.000	0.000	0.000	1.000

LAMBDA Y

	7.ERR.6	8.ERR_7	9.ERR_8	10.ERR_9	11.ERR10	12.ERR11
WISC_6	0.000	0.000	0.000	0.000	0.000	0.000
WISC_7	0.000	0.000	0.000	0.000	0.000	0.000
WISC_9	0.000	0.000	0.000	0.000	0.000	0.000
WISC_11	0.000	0.000	0.000	0.000	0.000	0.000

BETA

	1.WISC_6	2.WISC_7	3.WISC_8	4.WISC_9	5.WISC10	6.WISC11
1.WISC_6	0.000	0.000	0.000	0.000	0.000	0.000
2.WISC_7	0.987	0.000	0.000	0.000	0.000	0.000
3.WISC_8	0.000	0.987	0.000	0.000	0.000	0.000
4.WISC_9	0.000	0.000	0.987	0.000	0.000	0.000
5.WISC10	0.000	0.000	0.000	0.987	0.000	0.000
6.WISC11	0.000	0.000	0.000	0.000	0.987	0.000
7.ERR.6	0.000	0.000	0.000	0.000	0.000	0.000
8.ERR_7	0.000	0.000	0.000	0.000	0.000	0.000
9.ERR_8	0.000	0.000	0.000	0.000	0.000	0.000
10.ERR_9	0.000	0.000	0.000	0.000	0.000	0.000
11.ERR10	0.000	0.000	0.000	0.000	0.000	0.000
12.ERR11	0.000	0.000	0.000	0.000	0.000	0.000

BETA

	7.ERR.6	8.ERR_7	9.ERR_8	10.ERR_9	11.ERR10	12.ERR11
1.WISC_6	1.000	0.000	0.000	0.000	0.000	0.000
2.WISC_7	0.000	3.754	0.000	0.000	0.000	0.000
3.WISC_8	0.000	0.000	3.754	0.000	0.000	0.000
4.WISC_9	0.000	0.000	0.000	3.754	0.000	0.000
5.WISC10	0.000	0.000	0.000	0.000	3.754	0.000
6.WISC11	0.000	0.000	0.000	0.000	0.000	3.754
7.ERR.6	0.000	0.000	0.000	0.000	0.000	0.000
8.ERR_7	0.000	0.000	0.000	0.000	0.000	0.000
9.ERR_8	0.000	0.000	0.000	0.000	0.000	0.000
10.ERR_9	0.000	0.000	0.000	0.000	0.000	0.000
11.ERR10	0.000	0.000	0.000	0.000	0.000	0.000
12.ERR11	0.000	0.000	0.000	0.000	0.000	0.000

PSI

	1.WISC_6	2.WISC_7	3.WISC_8	4.WISC_9	5.WISC10	6.WISC11
1.WISC_6	44.347					
2 WISC_7	0.000	0.000				
3.WISC_8	0.000	0.000	0.000			
4.WISC_9	0.000	0.000	0.000	0.000		
5.WISC10	0.000	0.000	0.000	0.000	0.000	
6.WISC11	0.000	0.000	0.000	0.000	0.000	0.000
7.ERR.6	0.000	0.000	0.000	0.000	0.000	0.000
8.ERR_7	0.000	0.000	0.000	0.000	0.000	0.000
9.ERR_8	0.000	0.000	0.000	0.000	0.000	0.000
10.ERR_9	0.000	0.000	0.000	0.000	0.000	0.000
11.ERR10	0.000	0.000	0.000	0.000	0.000	0.000
12.ERR11	0.000	0.000	0.000	0.000	0.000	0.000

PSI

	7.ERR.6	8.ERR_7	9.ERR_8	10.ERR_9	11.ERR10	12.ERR11
7.ERR.6	1.000					
8.ERR_7	0.000	1.000				
9.ERR_8	0.000	0.000	1.000			
10.ERR_9	0.000	0.000	0.000	1.000		
11.ERR10	0.000	0.000	0.000	0.000	1.000	
12.ERR11	0.000	0.000	0.000	0.000	0.000	1.000

W_A_R_N_I_N_G : THE MATRIX PSI   IS NOT POSITIVE DEFINITE

MEASURES OF GOODNESS OF FIT FOR THE WHOLE MODEL :
CHI-SQUARE WITH   7 DEGREES OF FREEDOM IS   115.84 (PROB. LEVEL = 0.000)
GOODNESS OF FIT INDEX IS 0.802
ADJUSTED GOODNESS OF FIT INDEX IS 0.339
ROOT MEAN SQUARE RESIDUAL IS   9.556

The path coefficient estimates appear in the **BETA** matrix under the columns representing the WISC variables (both observed and unobserved). The residual standard deviations also appear in the **BETA** matrix under the **ERR__** variable columns. As mentioned above, the LISREL user familiar with Jöreskog and Sörbom's (1989) diagrams would expect to see those standard deviations appear as variances in the **PSI** and **THETA EPSILON** matrices. An equivalent model could be run estimating the variances in those locations.

Fit indices for the model follow the parameter estimates. These are followed by output normally used to assess the adequacy of the model. Although our input file included requests for modification indices, standard errors, a plot of normalized residuals against normal quantiles (Hoaglin *et al.* 1983) and a standardized solution, we list only *t*-values (testing the nontriviality of the estimates) and fitted moments, fitted residuals, and normalized residuals (showing how well the population model accounted for the covariances in the observed data (See McArdle & Aber (1990) for a comparison of a different autoregressive model on the same data).

T-VALUES

LAMBDA Y

	1.WISC_6	2.WISC_7	3.WISC_8	4.WISC_9	5.WISC10	6.WISC11
WISC_6	0.000	0.000	0.000	0.000	0.000	0.000
WISC_7	0.000	0.000	0.000	0.000	0.000	0.000
WISC_9	0.000	0.000	0.000	0.000	0.000	0.000
WISC_11	0.000	0.000	0.000	0.000	0.000	0.000

LAMBDA Y

	7.ERR.6	8.ERR_7	9.ERR_8	10.ERR_9	11.ERR10	12.ERR11
WISC_6	0.000	0.000	0.000	0.000	0.000	0.000
WISC_7	0.000	0.000	0.000	0.000	0.000	0.000
WISC_9	0.000	0.000	0.000	0.000	0.000	0.000
WISC_11	0.000	0.000	0.000	0.000	0.000	0.000

BETA

	1.WISC_6	2.WISC_7	3.WISC_8	4.WISC_9	5.WISC10	6.WISC11
1.WISC_6	0.000	0.000	0.000	0.000	0.000	0.000
2.WISC_7	67.923	0.000	0.000	0.000	0.000	0.000
3.WISC_8	0.000	67.923	0.000	0.000	0.000	0.000
4.WISC_9	0.000	0.000	67.923	0.000	0.000	0.000
5.WISC10	0.000	0.000	0.000	67.923	0.000	0.000
6.WISC11	0.000	0.000	0.000	0.000	67.923	0.000
7.ERR.6	0.000	0.000	0.000	0.000	0.000	0.000
8.ERR_7	0.000	0.000	0.000	0.000	0.000	0.000
9.ERR_8	0.000	0.000	0.000	0.000	0.000	0.000
10.ERR_9	0.000	0.000	0.000	0.000	0.000	0.000
11.ERR10	0.000	0.000	0.000	0.000	0.000	0.000
12.ERR11	0.000	0.000	0.000	0.000	0.000	0.000

BETA

	7.ERR.6	8.ERR_7	9.ERR_8	10.ERR_9	11.ERR10	12.ERR11
1.WISC_6	0.000	0.000	0.000	0.000	0.000	0.000
2.WISC_7	0.000	34.412	0.000	0.000	0.000	0.000
3.WISC_8	0.000	0.000	34.412	0.000	0.000	0.000
4.WISC_9	0.000	0.000	0.000	34.412	0.000	0.000
5.WISC10	0.000	0.000	0.000	0.000	34.412	0.000
6.WISC11	0.000	0.000	0.000	0.000	0.000	34.412
7.ERR.6	0.000	0.000	0.000	0.000	0.000	0.000
8.ERR_7	0.000	0.000	0.000	0.000	0.000	0.000
9.ERR_8	0.000	0.000	0.000	0.000	0.000	0.000
10.ERR_9	0.000	0.000	0.000	0.000	0.000	0.000
11.ERR10	0.000	0.000	0.000	0.000	0.000	0.000
12.ERR11	0.000	0.000	0.000	0.000	0.000	0.000

```
        PSI

            1.WISC_6   2.WISC_7   3.WISC_8   4.WISC_9   5.WISC10   6.WISC11

1.WISC_6       9.853
2.WISC_7       0.000      0.000
3.WISC_8       0.000      0.000      0.000
4.WISC_9       0.000      0.000      0.000      0.000
5.WISC10       0.000      0.000      0.000      0.000      0.000
6.WISC11       0.000      0.000      0.000      0.000      0.000      0.000
7.ERR.6        0.000      0.000      0.000      0.000      0.000      0.000
8.ERR_7        0.000      0.000      0.000      0.000      0.000      0.000
9.ERR_8        0.000      0.000      0.000      0.000      0.000      0.000
10.ERR_9       0.000      0.000      0.000      0.000      0.000      0.000
11.ERR10       0.000      0.000      0.000      0.000      0.000      0.000
12.ERR11       0.000      0.000      0.000      0.000      0.000      0.000
        PSI

            7.ERR.6    8.ERR_7    9.ERR_8   10.ERR_9   11.ERR10   12.ERR11

7.ERR.6        0.000
8.ERR_7        0.000      0.000
9.ERR_8        0.000      0.000      0.000
10.ERR_9       0.000      0.000      0.000      0.000
11.ERR10       0.000      0.000      0.000      0.000      0.000
12.ERR11       0.000      0.000      0.000      0.000      0.000      0.000

FITTED MOMENTS AND RESIDUALS

    FITTED MOMENTS

            WISC_6     WISC_7     WISC_9     WISC_11

WISC_6        45.347
WISC_7        44.750     58.256
WISC_9        43.579     56.732     83.069
WISC_11       42.439     55.247     80.896    106.602

    FITTED RESIDUALS

            WISC_6     WISC_7     WISC_9     WISC_11

WISC_6         0.000
WISC_7        -4.877     -4.688
WISC_9        -1.265     -8.232    -22.292
WISC_11       11.065      7.921    -10.696      1.267

    NORMALIZED RESIDUALS

            WISC_6     WISC_7     WISC_9     WISC_11

WISC_6         0.000
WISC_7        -1.020     -0.811
WISC_9        -0.240     -1.307     -2.704
WISC_11        1.935      1.173     -1.228      0.120
```

The normalized residuals are particularly helpful in locating those relationships in the data not accounted for by the model.

As is often the case, different indications of fit tell different stories. This model has many latent variables connected by many paths explaining very few covariances. The values of most of the paths are neccesarily constrained. The accompanying modification indices very often suggest that freely estimating a constrained parameter would improve the fit. Most of these suggestions represent either logically impossible (you could not freely estimate the basis of a linear growth curve; you cannot have time 2 causing time 1) or theoretically innappropriate changes to the model. In this model the chi-square test and the goodness-of-fit indices suggest that the covariances are not well accounted for, while the fitted and normalized residuals show there is not much left to fit in the matrix of associations. As McArdle and Aber (1990) point out, these fit indices may be best used to talk about relative fit in the context of a system of models, and they provide a discussion of relative fit for the different kinds of SEM models discussed here.

THE MOVING AVERAGE DIFFERENCE MODEL

Step 1: Preparation of the Data

The setup is the same as for the autoregression model and will result in the analysis of a covariance matrix of the variables.

Step 2: The Program

The LISREL program for computing the model follows:

```
USERPROC NAME=LISREL
TITLE MOVING AVERAGE DIFFERENCE MODEL
DA NG=1 NI=6 NOBS=204 MA=CM
LABELS
'WISC_6' 'WISC_7' 'WISC_9' 'WISC_11' 'MOTHERED' 'CONSTANT'
ME
    18.034 25.819 35.255 46.593 10.811 1.000
SD
    6.734 7.319 7.796 10.386 2.700 0.000
KM SY
*
1.000
 .809 1.000
 .806  .850 1.000
 .765  .831  .867 1.000
 .520  .531  .448  .458 1.000
 .000  .000  .000  .000  .000 1.000
SE
 1 2 3 4/
MO NY=4 NE=12 BE=FI,FU PS=FI,SY LY=FI TE=ZE
LE
   '1.WISC_6' '2.WISC_7' '3.WISC_9' '4.WISC11'
   '5.D_6' '6.D_7' '7.D_9' '8.D_11' 'ERR_6' 'ERR_7'
   'ERR_9' 'ERR_11'
MA LY
*
1 0 0 0 0 0 0 0 0 0 0 0
0 1 0 0 0 0 0 0 0 0 0 0
0 0 1 0 0 0 0 0 0 0 0 0
0 0 0 1 0 0 0 0 0 0 0 0
MA BE
*
0 0 0 0 1 0 0 0 1 0 0 0
0 0 0 0 1 1 0 0 0 1 0 0
0 0 0 0 1 1 0 0 0 1 0
0 0 0 0 0 1 1 0 0 0 1
```

```
0 0 0 0 0 0 0 0 0 0 0 0
0 0 0 0 0 0 0 0 0 0 0 0
0 0 0 0 0 0 0 0 0 0 0 0
0 0 0 0 0 0 0 0 0 0 0 0
0 0 0 0 0 0 0 0 0 0 0 0
0 0 0 0 0 0 0 0 0 0 0 0
0 0 0 0 0 0 0 0 0 0 0 0
0 0 0 0 0 0 0 0 0 0 0 0
MA PS
*
0
0 0
0 0 0
0 0 0 0
0 0 0 0 1
0 0 0 0 0 1
0 0 0 0 0 0 1
0 0 0 0 0 0 0 1
0 0 0 0 0 0 0 0 1
0 0 0 0 0 0 0 0 0 1
0 0 0 0 0 0 0 0 0 0 1
0 0 0 0 0 0 0 0 0 0 0 1
FR PS 9 9 PS 10 10 PS 11 11 PS 12 12
FR PS 5 5 PS 6 6 PS 7 7 PS 8 8
EQ PS 9 9 PS 10 10 PS 11 11 PS 12 12
OU PT SE TV NS MI SS RS ND=3 TO
END USER
```

The program setup is much the same as the autoregression program listed above. For variety's sake, the residuals variances are estimated as variances in the diagonals of the **PSI** matrix. An equivalent model would fix the **PSI** values to 1 and estimate those parameters as standard deviations by freeing the appropriate **BETA** values.

Step 3: Output for the Moving-Average Difference Model

LISREL estimates for this model follow:

```
TITLE MOVING AVERAGE DIFFERENCE MODEL

LISREL ESTIMATES (MAXIMUM LIKELIHOOD)
```

LAMBDA Y

	1.WISC_6	2.WISC_7	3.WISC_9	4.WISC11	5.D_6	6.D_7
WISC_6	1.000	0.000	0.000	0.000	0.000	0.000
WISC_7	0.000	1.000	0.000	0.000	0.000	0.000
WISC_9	0.000	0.000	1.000	0.000	0.000	0.000
WISC_11	0.000	0.000	0.000	1.000	0.000	0.000

LAMBDA Y

	7.D_9	8.D_11	ERR_6	ERR_7	ERR_9	ERR_11
WISC_6	0.000	0.000	0.000	0.000	0.000	0.000
WISC_7	0.000	0.000	0.000	0.000	0.000	0.000
WISC_9	0.000	0.000	0.000	0.000	0.000	0.000
WISC_11	0.000	0.000	0.000	0.000	0.000	0.000

BETA

	1.WISC_6	2.WISC_7	3.WISC_9	4.WISC11	5.D_6	6.D_7
1.WISC_6	0.000	0.000	0.000	0.000	1.000	0.000
2.WISC_7	0.000	0.000	0.000	0.000	1.000	1.000

3.WISC_9	0.000	0.000	0.000	0.000	0.000	1.000
4.WISC11	0.000	0.000	0.000	0.000	0.000	0.000
5.D_6	0.000	0.000	0.000	0.000	0.000	0.000
6.D_7	0.000	0.000	0.000	0.000	0.000	0.000
7.D_9	0.000	0.000	0.000	0.000	0.000	0.000
8.D_11	0.000	0.000	0.000	0.000	0.000	0.000
ERR_6	0.000	0.000	0.000	0.000	0.000	0.000
ERR_7	0.000	0.000	0.000	0.000	0.000	0.000
ERR_9	0.000	0.000	0.000	0.000	0.000	0.000
ERR_11	0.000	0.000	0.000	0.000	0.000	0.000

BETA

	7.D_9	8.D_11	ERR_6	ERR_7	ERR_9	ERR_11
1.WISC_6	0.000	0.000	1.000	0.000	0.000	0.000
2.WISC_7	0.000	0.000	0.000	1.000	0.000	0.000
3.WISC_9	1.000	0.000	0.000	0.000	1.000	0.000
4.WISC11	1.000	1.000	0.000	0.000	0.000	1.000
5.D_6	0.000	0.000	0.000	0.000	0.000	0.000
6.D_7	0.000	0.000	0.000	0.000	0.000	0.000
7.D_9	0.000	0.000	0.000	0.000	0.000	0.000
8.D_11	0.000	0.000	0.000	0.000	0.000	0.000
ERR_6	0.000	0.000	0.000	0.000	0.000	0.000
ERR_7	0.000	0.000	0.000	0.000	0.000	0.000
ERR_9	0.000	0.000	0.000	0.000	0.000	0.000
ERR_11	0.000	0.000	0.000	0.000	0.000	0.000

PSI

	1.WISC_6	2.WISC_7	3.WISC_9	4.WISC11	5.D_6	6.D_7
1.WISC_6	0.000					
2.WISC_7	0.000	0.000				
3.WISC_9	0.000	0.000	0.000			
4.WISC11	0.000	0.000	0.000	0.000		
5.D_6	0.000	0.000	0.000	0.000	40.653	
6.D_7	0.000	0.000	0.000	0.000	0.000	-0.621
7.D_9	0.000	0.000	0.000	0.000	0.000	0.000
8.D_11	0.000	0.000	0.000	0.000	0.000	0.000
ERR_6	0.000	0.000	0.000	0.000	0.000	0.000
ERR_7	0.000	0.000	0.000	0.000	0.000	0.000
ERR_9	0.000	0.000	0.000	0.000	0.000	0.000
ERR_11	0.000	0.000	0.000	0.000	0.000	0.000

PSI

	7.D_9	8.D_11	ERR_6	ERR_7	ERR_9	ERR_11
7.D_9	62.367					
8.D_11	0.000	14.701				
ERR_6	0.000	0.000	8.834			
ERR_7	0.000	0.000	0.000	8.834		
ERR_9	0.000	0.000	0.000	0.000	8.834	
ERR_11	0.000	0.000	0.000	0.000	0.000	8.834

W_A_R_N_I_N_G : THE MATRIX PSI   IS NOT POSITIVE DEFINITE

SQUARED MULTIPLE CORRELATIONS FOR STRUCTURAL EQUATIONS

1.WISC_6	2.WISC_7	3.WISC_9	4.WISC11	5.D_6	6.D_7
1.000	1.000	1.000	1.000	0.000	0.000

SQUARED MULTIPLE CORRELATIONS FOR STRUCTURAL EQUATIONS

7.D_9	8.D_11	ERR_6	ERR_7	ERR_9	ERR_11
0.000	0.000	0.000	0.000	0.000	0.000

MEASURES OF GOODNESS OF FIT FOR THE WHOLE MODEL :
CHI-SQUARE WITH   5 DEGREES OF FREEDOM IS   361.85 (PROB. LEVEL = 0.000)
GOODNESS OF FIT INDEX IS 0.640
ADJUSTED GOODNESS OF FIT INDEX IS 0.280
ROOT MEAN SQUARE RESIDUAL IS   34.258

T-VALUES

PSI

	1.WISC_6	2.WISC_7	3.WISC_9	4.WISC11	5.D_6	6.D_7
1.WISC_6	0.000					
2.WISC_7	0.000	0.000				
3.WISC_9	0.000	0.000	0.000			
4.WISC11	0.000	0.000	0.000	0.000		
5.D_6	0.000	0.000	0.000	0.000	9.082	
6.D_7	0.000	0.000	0.000	0.000	0.000	-0.488
7.D_9	0.000	0.000	0.000	0.000	0.000	0.000
8.D_11	0.000	0.000	0.000	0.000	0.000	0.000
ERR_6	0.000	0.000	0.000	0.000	0.000	0.000
ERR_7	0.000	0.000	0.000	0.000	0.000	0.000
ERR_9	0.000	0.000	0.000	0.000	0.000	0.000
ERR_11	0.000	0.000	0.000	0.000	0.000	0.000

PSI

	7.D_9	8.D_11	ERR_6	ERR_7	ERR_9	ERR_11
7.D_9	9.149					
8.D_11	0.000	4.335				
ERR_6	0.000	0.000	8.124			
ERR_7	0.000	0.000	0.000	8.124		
ERR_9	0.000	0.000	0.000	0.000	8.124	
ERR_11	0.000	0.000	0.000	0.000	0.000	8.124

FITTED MOMENTS AND RESIDUALS

FITTED MOMENTS

	WISC_6	WISC_7	WISC_9	WISC_11
WISC_6	49.488			
WISC_7	40.653	48.866		
WISC_9	0.000	-0.621	70.580	
WISC_11	0.000	0.000	62.367	85.902

FITTED RESIDUALS

	WISC_6	WISC_7	WISC_9	WISC_11
WISC_6	-4.141			
WISC_7	-0.781	4.701		
WISC_9	42.314	49.121	-9.802	
WISC_11	53.504	63.169	7.834	21.967

NORMALIZED RESIDUALS

	WISC_6	WISC_7	WISC_9	WISC_11
WISC_6	-0.843			
WISC_7	-0.174	0.969		
WISC_9	10.201	11.916	-1.399	
WISC_11	11.692	13.891	1.119	2.576

Once again, the **LAMBDA** matrix defines four of the **ETA** variables as observed. The **BETA** matrix reflects the fixed (to 1) values of the path coefficient defining the difference conponents. In this model the **PSI** matrix holds the values of the residual variances for the four difference components and for the observed variables. This solution includes a Heywood case (van Driel, 1978). The residual variance for **D__7** is negative, but very small. If this value repesents sampling fluctuation around a true value of 0, the parameter estimate could be fixed to 0. If this value were large and negative, it would probably indicate something fundamentally wrong with the model.

The estimates are followed by the fit indices. Again, we only include part of the requested output. We also urge caution and suggest the user not use the goodness-of-fit indices as the *only* consideration regarding whether the model is good or bad.

Information regarding output options and interpretations is plentiful (Jöreskog & Sörbom, 1989; Hayduk, 1987; Bollen, 1989; Long, 1983).

THE LINEAR LATENT GROWTH-CURVE MODEL

Step 1:   Preparation of the Data

This model will be estimating mean levels of the latent variables in addition to parameters describing associations in the data. This cannot be accomplished with a covariance matrix. The problem requires an input matrix of cross-products. Since the mean of a variable can be estimated by regressing that variable onto a constant, a constant must be included as an input variable. One way to request this input is to use the augmented moment matrix of LISREL (**MA=AM**). If you provide the program with a correlation matrix and vectors of means and standard deviations, it will compute the moment matrix and add a variable called 'CONST' to the data set. By including 'CONST' as one of the selected variables, you will have the augmented matrix available. An alternative specification would have the user read in the augmented moment matrix but call the matrix a covariance matrix (**MA=CM**). (WARNING: If you read in the moment matrix and call it a moment matrix (**MA MM**), LISREL will do some very strange rescaling.)

Step 2:   The Program (Adapted from McArdle and Epstein (1987)).

The input for the linear latent-growth-curve model follows:

```
USERPROC NAME=LISREL
TITLE LINEAR LATENT GROWTH CURVE MODEL
DA NG=1 NI=4 NOBS=204 MA=AM
LABELS
*
'WISC_6' 'WISC_7' 'WISC_9' 'WISC_11'
ME
    18.034 25.819 35.255 46.593
SD
    6.734 7.319 7.796 10.386
KM SY
*
1.000
 .809 1.000
 .806  .850 1.000
 .765  .831  .867 1.000
SE
 1 2 3 4 'CONST'/
MO NY=5 NE=13 BE=FI,FU PS=FI,SY LY=FI TE=ZE
```

```
LE
   '1.WISC_6' '2.WISC_7' '3.WISC_9' '4.WISC11'
   '5.ERR_6' '6.ERR_7' '7.ERR_9' '8.ERR11'
   '9.LEVEL' '10.SHAPE' 'CONST' '12.STD.L' '13.STD.S'
MA LY
*
1 0 0 0 0 0 0 0 0 0 0 0 0
0 1 0 0 0 0 0 0 0 0 0 0 0
0 0 1 0 0 0 0 0 0 0 0 0 0
0 0 0 1 0 0 0 0 0 0 0 0 0
0 0 0 0 0 0 0 0 0 1 0 0
MA BE
*
0 0 0 0 1 0 0 0 1 0.000 0 0 0
0 0 0 0 0 1 0 0 1 0.186 0 0 0
0 0 0 0 0 0 1 0 1 0.579 0  0 0
0 0 0 0 0 0 1 1 1.000 0  0 0
0 0 0 0 0 0 0 0 0 0  0 0
0 0 0 0 0 0 0 0 0 0  0 0
0 0 0 0 0 0 0 0 0 0  0 0
0 0 0 0 0 0 0 0 0 0  0 0
0 0 0 0 0 0 0 0 0 1  1 0
0 0 0 0 0 0 0 0 0 1  0 1
0 0 0 0 0 0 0 0 0 0  0 0
0 0 0 0 0 0 0 0 0 0  0 0
0 0 0 0 0 0 0 0 0 0  0 0
MA PS
*
0
0 0
0 0 0
0 0 0 0
0 0 0 0 1
0 0 0 0 0 1
0 0 0 0 0 0 1
0 0 0 0 0 0 0 1
0 0 0 0 0 0 0 0 0
0 0 0 0 0 0 0 0 0 0
0 0 0 0 0 0 0 0 0 1
0 0 0 0 0 0 0 0 0 0 1
0 0 0 0 0 0 0 0 0 0 .5 1
FR PS 5 5 PS 6 6 PS 7 7 PS 8 8 PS 12 12 PS 13 13 PS 13 12
EQ PS 5 5 PS 6 6 PS 7 7 PS 8 8
FR BE 9 11 BE 10 11
OU PT SE TV NS MI SS RS NO=3 TO
END USER
```

The four observed variables are defined and followed in the **BETA** matrix by the four residual variables associated with each of the observed variables respectively. The latent variables, Level and Shape, account for some of the covariation among the observed variables. In this model the values in **BETA** for the paths from **L** to the observed variables are fixed to 1. The values for the paths from **S** to the observed variables form the basis vector and either define (if they are fixed) the type of curve fit to the data or estimate (if they are free) the shape of the curve. In this case the basis is fixed and represents a linear curve for unequally spaced observations. Estimates of the means of **L** and **S** appear as the **BETA** matrix as **BETA 9 11** and **BETA 10 11**. The covariance between **L** and **S** appears in **PSI 13 12**. In this model residual variances are estimated in the **PSI** matrix with the **PSI 11 11** value fixed to

1. This in principle represents the fact that none of the **CONST**'s variance (even though it has none) is being estimated by anything else in the model. As the constant is included in the input matrix, it should be thought of primarily as a variable that correlates 0 with all other variables in the model.

Step 3: Output for the Linear Growth-Curve Model

Because this model may be new to many familiar with SEMs, we will include relatively more of the output from this model run.

The input to this model included a correlation matrix along with means and standard deviations for the observed variables along with a request for the augmented moment matrix. This produced the following matrix to be analyzed:

MOMENT MATRIX TO BE ANALYZED

	WISC_6	WISC_7	WISC_9	WISC_11	CONST.
WISC_6	370.350				
WISC_7	505.297	719.926			
WISC_9	677.895	958.511	1303.395		
WISC_11	893.499	1265.844	1712.492	2278.248	
CONST.	18.034	25.819	35.255	46.593	1.000

DETERMINANT = 0.274176D+06

The bottom row relating the constant to each variable produces the means of the input variables.

The parameter specifications and starting values for this model are

TITLE LINEAR LATENT GROWTH CURVE MODEL

PARAMETER SPECIFICATIONS

LAMBDA Y

	1.WISC_6	2.WISC_7	3.WISC_9	4.WISC11	5.ERR_6	6.ERR_7
WISC_6	0	0	0	0	0	0
WISC_7	0	0	0	0	0	0
WISC_9	0	0	0	0	0	0
WISC_11	0	0	0	0	0	0
CONST.	0	0	0	0	0	0

LAMBDA Y

	7.ERR_9	8.ERR11	9.LEVEL	10.SHAPE	CONST	12.STD.L
WISC_6	0	0	0	0	0	0
WISC_7	0	0	0	0	0	0
WISC_9	0	0	0	0	0	0
WISC_11	0	0	0	0	0	0
CONST.	0	0	0	0	0	0

LAMBDA Y

	13.STD.S
WISC_6	0
WISC_7	0
WISC_9	0
WISC_11	0
CONST.	0

BETA

	1.WISC_6	2.WISC_7	3.WISC_9	4.WISC11	5.ERR_6	6.ERR_7
1.WISC_6	0	0	0	0	0	0
2.WISC_7	0	0	0	0	0	0
3.WISC_9	0	0	0	0	0	0
4.WISC11	0	0	0	0	0	0
5.ERR_6	0	0	0	0	0	0
6.ERR_7	0	0	0	0	0	0
7.ERR_9	0	0	0	0	0	0
8.ERR11	0	0	0	0	0	0
9.LEVEL	0	0	0	0	0	0
10.SHAPE	0	0	0	0	0	0
CONST	0	0	0	0	0	0
12.STD.L	0	0	0	0	0	0
13.STD.S	0	0	0	0	0	0

BETA

	7.ERR_9	8.ERR11	9.LEVEL	10.SHAPE	CONST	12.STD.L
1.WISC_6	0	0	0	0	0	0
2.WISC_7	0	0	0	0	0	0
3.WISC_9	0	0	0	0	0	0
4.WISC11	0	0	0	0	0	0
5.ERR_6	0	0	0	0	0	0
6.ERR_7	0	0	0	0	0	0
7.ERR_9	0	0	0	0	0	0
8.ERR11	0	0	0	0	0	0
9.LEVEL	0	0	0	0	1	0
10.SHAPE	0	0	0	0	2	0
CONST	0	0	0	0	0	0
12.STD.L	0	0	0	0	0	0
13.STD.S	0	0	0	0	0	0

BETA

	13.STD.S
1.WISC_6	0
2.WISC_7	0
3.WISC_9	0
4.WISC11	0
5.ERR_6	0
6.ERR_7	0
7.ERR_9	0
8.ERR11	0
9.LEVEL	0
10.SHAPE	0
CONST	0
12.STD.L	0
13.STD.S	0

PSI

	1.WISC_6	2.WISC_7	3.WISC_9	4.WISC11	5.ERR_6	6.ERR_7
1.WISC_6	0					
2.WISC_7	0	0				
3.WISC_9	0	0	0			
4.WISC11	0	0	0	0		
5.ERR_6	0	0	0	0	3	

6.ERR_7	0	0	0	0	0	3
7.ERR_9	0	0	0	0	0	0
8.ERR11	0	0	0	0	0	0
9.LEVEL	0	0	0	0	0	0
10.SHAPE	0	0	0	0	0	0
CONST	0	0	0	0	0	0
12.STD.L	0	0	0	0	0	0
13.STD.S	0	0	0	0	0	0

PSI

	7.ERR_9	8.ERR11	9.LEVEL	10.SHAPE	CONST	12.STD.L
7.ERR_9	3					
8.ERR11	0	3				
9.LEVEL	0	0	0			
10.SHAPE	0	0	0	0		
CONST	0	0	0	0	0	
12.STD.L	0	0	0	0	0	4
13.STD.S	0	0	0	0	0	5

PSI

	13.STD.S
13.STD.S	6

STARTING VALUES

LAMBDA Y

	1.WISC_6	2.WISC_7	3.WISC_9	4.WISC11	5.ERR_6	6.ERR_7
WISC_6	1.000	0.000	0.000	0.000	0.000	0.000
WISC_7	0.000	1.000	0.000	0.000	0.000	0.000
WISC_9	0.000	0.000	1.000	0.000	0.000	0.000
WISC_11	0.000	0.000	0.000	1.000	0.000	0.000
CONST.	0.000	0.000	0.000	0.000	0.000	0.000

LAMBDA Y

	7.ERR_9	8.ERR11	9.LEVEL	10.SHAPE	CONST	12.STD.L
WISC_6	0.000	0.000	0.000	0.000	0.000	0.000
WISC_7	0.000	0.000	0.000	0.000	0.000	0.000
WISC_9	0.000	0.000	0.000	0.000	0.000	0.000
WISC_11	0.000	0.000	0.000	0.000	0.000	0.000
CONST.	0.000	0.000	0.000	0.000	1.000	0.000

LAMBDA Y

	13.STD.S
WISC_6	0.000
WISC_7	0.000
WISC_9	0.000
WISC_11	0.000
CONST.	0.000

BETA

	1.WISC_6	2.WISC_7	3.WISC_9	4.WISC11	5.ERR_6	6.ERR_7
1.WISC_6	0.000	0.000	0.000	0.000	1.000	0.000
2.WISC_7	0.000	0.000	0.000	0.000	0.000	1.000
3.WISC_9	0.000	0.000	0.000	0.000	0.000	0.000
4.WISC11	0.000	0.000	0.000	0.000	0.000	0.000
5.ERR_6	0.000	0.000	0.000	0.000	0.000	0.000
6.ERR_7	0.000	0.000	0.000	0.000	0.000	0.000
7.ERR_9	0.000	0.000	0.000	0.000	0.000	0.000
8.ERR11	0.000	0.000	0.000	0.000	0.000	0.000
9.LEVEL	0.000	0.000	0.000	0.000	0.000	0.000
10.SHAPE	0.000	0.000	0.000	0.000	0.000	0.000
CONST	0.000	0.000	0.000	0.000	0.000	0.000
12.STD.L	0.000	0.000	0.000	0.000	0.000	0.000
13.STD.S	0.000	0.000	0.000	0.000	0.000	0.000

BETA

	7.ERR_9	8.ERR11	9.LEVEL	10.SHAPE	CONST	12.STD.L
1.WISC_6	0.000	0.000	1.000	0.000	0.000	0.000
2.WISC_7	0.000	0.000	1.000	0.186	0.000	0.000
3.WISC_9	1.000	0.000	1.000	0.579	0.000	0.000
4.WISC11	0.000	1.000	1.000	1.000	0.000	0.000
5.ERR_5	0.000	0.000	0.000	0.000	0.000	0.000
6.ERR_7	0.000	0.000	0.000	0.000	0.000	0.000
7.ERR_9	0.000	0.000	0.000	0.000	0.000	0.000
8.ERR11	0.000	0.000	0.000	0.000	0.000	0.000
9.LEVEL	0.000	0.000	0.000	0.000	1.000	1.000
10.SHAPE	0.000	0.000	0.000	0.000	1.000	0.000
CONST	0.000	0.000	0.000	0.000	0.000	0.000
12.STD.L	0.000	0.000	0.000	0.000	0.000	0.000
13.STD.S	0.000	0.000	0.000	0.000	0.000	0.000

BETA

	13.STD.S
1.WISC_6	0.000
2.WISC_7	0.000
3.WISC_9	0.000
4.WISC11	0.000
5.ERR_6	0.000
6.ERR_7	0.000
7.ERR_9	0.000
8.ERR11	0.000
9.LEVEL	0.000
10.SHAPE	1.000
CONST	0.000
12.STD.L	0.000
13.STD.S	0.000

PSI

	1.WISC_6	2.WISC_7	3.WISC_9	4.WISC11	5.ERR_6	6.ERR_7
1.WISC_6	0.000					
2.WISC_7	0.000	0.000				
3.WISC_9	0.000	0.000	0.000			
4.WISC11	0.000	0.000	0.000	0.000		
5.ERR_6	0.000	0.000	0.000	0.000	1.000	
6.ERR_7	0.000	0.000	0.000	0.000	0.000	1.000
7.ERR_9	0.000	0.000	0.000	0.000	0.000	0.000
8.ERR11	0.000	0.000	0.000	0.000	0.000	0.000
9.LEVEL	0.000	0.000	0.000	0.000	0.000	0.000
10.SHAPE	0.000	0.000	0.000	0.000	0.000	0.000
CONST	0.000	0.000	0.000	0.000	0.000	0.000
12.STD.L	0.000	0.000	0.000	0.000	0.000	0.000
13.STD.S	0.000	0.000	0.000	0.000	0.000	0.000

PSI

	7.ERR_9	8.ERR11	9.LEVEL	10.SHAPE	CONST	12.STD.L
7.ERR_9	1.000					
8.ERR11	0.000	1.000				
9.LEVEL	0.000	0.000	0.000			
10.SHAPE	0.000	0.000	0.000	0.000		
CONST	0.000	0.000	0.000	0.000	1.000	
12.STD.L	0.000	0.000	0.000	0.000	0.000	1.000
13.STD.S	0.000	0.000	0.000	0.000	0.000	0.500

PSI

	13.STD.S
13.STD.S	1.000

In addition to the observed variables and their residuals, we have latent variables of Level and Shape along with external variables contributing to their covariation. Here they are called **Ext.L** and **Ext.S**, and we estimate their covariance (**PSI 13 12**). Six parameters are to be estimated in this model.

The starting values include the fixed basis vector b_i (McArdle and Aber, 1990) for a linear curve with unequal spacing. This is one of many possible fixed vectors we could use to look for a specific curve. Another option would be to freely estimate the basis vector to determine the shape of the curve.

The LISREL estimates for this model are

LISREL ESTIMATES (MAXIMUM LIKELIHOOD)

LAMBDA Y

	1.WISC_6	2.WISC_7	3.WISC_9	4.WISC11	5.ERR_6	6.ERR_7
WISC_6	1.000	0.000	0.000	0.000	0.000	0.000
WISC_7	0.000	1.000	0.000	0.000	0.000	0.000
WISC_9	0.000	0.000	1.000	0.000	0.000	0.000
WISC_11	0.000	0.000	0.000	1.000	0.000	0.000
CONST.	0.000	0.000	0.000	0.000	0.000	0.000

LAMBDA Y

	7.ERR_9	8.ERR11	9.LEVEL	10.SHAPE	CONST	12.STD.L
WISC_6	0.000	0.000	0.000	0.000	0.000	0.000
WISC_7	0.000	0.000	0.000	0.000	0.000	0.000
WISC_9	0.000	0.000	0.000	0.000	0.000	0.000
WISC_11	0.000	0.000	0.000	0.000	0.000	0.000
CONST.	0.000	0.000	0.000	0.000	1.000	0.000

LAMBDA Y

	13.STD.S
WISC_6	0.000
WISC_7	0.000
WISC_9	0.000
WISC_11	0.000
CONST.	0.000

BETA

	1.WISC_6	2.WISC_7	3.WISC_9	4.WISC11	5.ERR_6	6.ERR_7
1.WISC_6	0.000	0.000	0.000	0.000	1.000	0.000
2.WISC_7	0.000	0.000	0.000	0.000	0.000	1.000
3.WISC_9	0.000	0.000	0.000	0.000	0.000	0.000
4.WISC11	0.000	0.000	0.000	0.000	0.000	0.000
5.ERR_6	0.000	0.000	0.000	0.000	0.000	0.000
6.ERR_7	0.000	0.000	0.000	0.000	0.000	0.000
7.ERR_9	0.000	0.000	0.000	0.000	0.000	0.000
8.ERR11	0.000	0.000	0.000	0.000	0.000	0.000
9.LEVEL	0.000	0.000	0.000	0.000	0.000	0.000
10.SHAPE	0.000	0.000	0.000	0.000	0.000	0.000
CONST	0.000	0.000	0.000	0.000	0.000	0.000
12.STD.L	0.000	0.000	0.000	0.000	0.000	0.000
13.STD.S	0.000	0.000	0.000	0.000	0.000	0.000

BETA

	7.ERR_9	8.ERR11	9.LEVEL	10.SHAPE	CONST	12.STD.L
1.WISC_6	0.000	0.000	1.000	0.000	0.000	0.000
2.WISC_7	0.000	0.000	1.000	0.186	0.000	0.000

3.WISC_9	1.000	0.000	1.000	0.579	0.000	0.000
4.WISC11	0.000	1.000	1.000	1.000	0.000	0.000
5.ERR_6	0.000	0.000	0.000	0.000	0.000	0.000
6.ERR_7	0.000	0.000	0.000	0.000	0.000	0.000
7.ERR_9	0.000	0.000	0.000	0.000	0.000	0.000
8.ERR11	0.000	0.000	0.000	0.000	0.000	0.000
9.LEVEL	0.000	0.000	0.000	0.000	19.224	1.000
10.SHAPE	0.000	0.000	0.000	0.000	27.651	0.000
CONST	0.000	0.000	0.000	0.000	0.000	0.000
12.STD.L	0.000	0.000	0.000	0.000	0.000	0.000
13.STD.S	0.000	0.000	0.000	0.000	0.000	0.000

        BETA

	13.STD.S
1.WISC_6	0.000
2.WISC_7	0.000
3.WISC_9	0.000
4.WISC11	0.000
5.ERR_6	0.000
6.ERR_7	0.000
7.ERR_9	0.000
8.ERR11	0.000
9.LEVEL	0.000
10.SHAPE	1.000
CONST	0.000
12.STD.L	0.000
13.STD.S	0.000

        PSI

	1.WISC_6	2.WISC_7	3.WISC_9	4.WISC11	5.ERR_6	6.ERR_7
1.WISC_6	0.000					
2.WISC_7	0.000	0.000				
3.WISC_9	0.000	0.000	0.000			
4.WISC11	0.000	0.000	0.000	0.000		
5.ERR_6	0.000	0.000	0.000	0.000	10.875	
6.ERR_7	0.000	0.000	0.000	0.000	0.000	10.875
7.ERR_9	0.000	0.000	0.000	0.000	0.000	0.000
8.ERR11	0.000	0.000	0.000	0.000	0.000	0.000
9.LEVEL	0.000	0.000	0.000	0.000	0.000	0.000
10.SHAPE	0.000	0.000	0.000	0.000	0.000	0.000
CONST	0.000	0.000	0.000	0.000	0.000	0.000
12.STD.L	0.000	0.000	0.000	0.000	0.000	0.000
13.STD.S	0.000	0.000	0.000	0.000	0.000	0.000

        PSI

	7.ERR_9	8.ERR11	9.LEVEL	10.SHAPE	CONST	12.STD.L
7.ERR_9	10.875					
8.ERR11	0.000	10.875				
9.LEVEL	0.000	0.000	0.000			
10.SHAPE	0.000	0.000	0.000	0.000		
CONST	0.000	0.000	0.000	0.000	1.000	
12.STD.L	0.000	0.000	0.000	0.000	0.000	34.426
13.STD.S	0.000	0.000	0.000	0.000	0.000	17.211

        PSI

	13.STD.S
13.STD.S	21.525

MEASURES OF GOODNESS OF FIT FOR THE WHOLE MODEL :
CHI-SQUARE WITH   9 DEGREES OF FREEDOM IS   90.67 (PROB. LEVEL = 0.000)
GOODNESS OF FIT INDEX IS 0.855
ADJUSTED GOODNESS OF FIT INDEX IS 0.639
ROOT MEAN SQUARE RESIDUAL IS   36.744

BETA 9 11 and **10 11** estimate the intercept and the slope, respectively. Note that, as often happens in regression, the intercept does not equal the mean. We could make the intercept a test of the mean by centering the basis (generating an equivalent basis with mean=0). With four equally spaced occasions one could use linear polynomial coefficients $(-3/sqrt(20), -1/sqrt(20), 1/sqrt(20), 3/sqrt(20))$. This basis would test both linearity and the grand mean of the WISC data.

A variance/covariance matrix of the latent variables sits in the lower right corner of the **PSI** matrix.

Modification indices generated by this model are weird and are provided primarily for the reader's entertainment. They, in part, represent the attempt to calculate second derivatives using a variable with no variance. They are probably not appropriate for this kind of model.

```
MODIFICATION INDICES

        LAMBDA Y

            1.WISC_6   2.WISC_7   3.WISC_9   4.WISC11   5.ERR_6    6.ERR_7

WISC_6       49.213     60.554     55.778     58.748     0.897      5.512
WISC_7       55.833     61.838     63.867     62.414     5.512      0.749
WISC_9        0.309      0.258      0.694      0.310     1.283      1.657
WISC_11       3.009      3.307      3.700      3.312     0.076      0.063
CONST.        0.466      0.014      1.063      0.144     4.410      0.017
        LAMBDA Y

            7.ERR_9    8.ERR11    9.LEVEL   10.SHAPE   CONST    12.STD.L

WISC_6        1.283      0.076     55.830     60.499    62.894     0.253
WISC_7        1.657      0.063     61.337     63.503    61.482     1.184
WISC_9        9.586      0.079      0.529      0.170     0.011     8.755
WISC_11       0.079      2.030      2.989      4.386     6.698     7.268
CONST.       13.167     14.322      0.007      0.001     0.000***********
        LAMBDA Y
        13.STD.S

WISC_6        0.269
WISC_7        2.137
WISC_9       13.433
WISC_11      11.001
CONST. ***********

FLOATING POINT DIVIDE CHECK (DIVISION BY ZERO) AT ADDRESS 7978DA
SUBROUTINE TRACEBACK - CURRENT ROUTINE FIRST
ROUTINE                    STMT     REG 14    REG 15     REG 0     REG 1
FMOD                                6078F79C  00797010  00115260  0073DB88

FLOATING POINT OVERFLOW (RESULT TOO LARGE) AT ADDRESS 797BE6
SUBROUTINE TRACEBACK - CURRENT ROUTINE FIRST
ROUTINE                    STMT     REG 14    REG 15     REG 0     REG 1
FMOD                                6078F79C  00797010  00115260  0073DB88

        BETA
            1.WISC_6   2.WISC_7   3.WISC_9   4.WISC11   5.ERR_6    6.ERR_7

1.WISC_6     49.213     60.554     55.778     58.748     0.897      5.512
2.WISC_7     55.833     61.838     63.867     62.414     5.512      0.749
```

3.WISC_9	0.309	0.258	0.694	0.310	1.283	1.657
4.WISC11	3.009	3.307	3.700	3.312	0.076	0.063
5.ERR_6	49.213	60.554	55.778	58.748	0.897	5.512
6.ERR_7	55.833	61.838	63.867	62.414	5.512	0.749
7.ERR_9	0.309	0.258	0.694	0.310	1.283	1.657
8.ERR11	3.009	3.307	3.700	3.312	0.076	0.063
9.LEVEL	0.024	0.071	3.013	0.897	0.128	0.204
10.SHAPE	0.041	1.440	9.175	5.752	0.091	1.664
CONST	0.387	0.358	0.000	0.004	62.894	61.482
12.STD.L	0.024	0.071	3.013	0.897	0.128	0.204
13.STD.S	0.041	1.440	9.175	5.752	0.091	1.664

BETA

	7.ERR_9	8.ERR11	9.LEVEL	10.SHAPE	CONST	12.STD.L
1.WISC_6	1.283	0.076	55.830	60.499	62.894	0.253
2.WISC_7	1.657	0.063	61.337	63.503	61.482	1.184
3.WISC_9	9.586	0.079	0.529	0.170	0.011	8.755
4.WISC11	0.079	2.030	2.989	4.386	6.698	7.268
5.ERR_6	1.283	0.076	55.830	60.499	62.894	0.253
6.ERR_7	1.657	0.063	61.337	63.503	61.482	1.184
7.ERR_9	9.586	0.079	0.529	0.170	0.011	8.755
8.ERR11	0.079	2.030	2.989	4.386	6.698	7.268
9.LEVEL	2.317	1.983	**********************		0.000	***********
10.SHAPE	8.556	6.878	**********************		0.000	***********
CONST .	0.011	6.698	0.000	0.000	0.000	0.000
12.STD.L	2.317	1.983	**********************		0.000	***********
13.STD.S	8.556	6.878	**********************		0.000	***********

BETA

	13.STD.S
1.WISC_6	0.269
2.WISC_7	2.137
3.WISC_9	13.433
4.WISC11	11.001
5.ERR_6	0.269
6.ERR_7	2.137
7.ERR_9	13.433
8.ERR11	11.001
9.LEVEL	***********
10.SHAPE	***********
CONST	0.000
12.STD.L	***********
13.STD.S	***********

PSI

	1.WISC_6	2.WISC_7	3.WISC_9	4.WISC11	5.ERR_6	6.ERR_7
1.WISC_6	0.897					
2.WISC_7	5.512	0.749				
3.WISC_9	1.283	1.657	9.586			
4.WISC11	0.076	0.063	0.079	2.030		
5.ERR_6	0.897	5.512	1.283	0.076	0.933	
6.ERR_7	5.512	0.749	1.657	0.063	5.512	0.760
7.ERR_9	1.283	1.657	9.586	0.079	1.283	1.657
8.ERR11	0.076	0.063	0.079	2.030	0.076	0.063
9.LEVEL	0.128	0.204	2.317	1.983	0.128	0.204
10.SHAPE	0.091	1.664	8.556	6.878	0.091	1.664
CONST	62.894	61.482	0.011	6.698	62.894	61.482
12.STD.L	0.128	0.204	2.317	1.983	0.128	0.204
13.STD.S	0.091	1.664	8.556	6.878	0.091	1.664

PSI

	7.ERR_9	8.ERR11	9.LEVEL	10.SHAPE	CONST	12.STD.L
7.ERR_9	10.877					
8.ERR11	0.079	2.339				
9.LEVEL	2.317	1.983	***********			

```
10.SHAPE        8.556      6.878**********      0.000
CONST           0.011      6.698      0.000     0.000      0.000
12.STD.L        2.317      1.983********************      0.000      0.000
13.STD.S        8.556      6.878**********      0.000      0.000      0.000
            PSI

                13.STD.S

13.STD.S        0.000

        MAXIMUM MODIFICATION INDEX IS******** FOR ELEMENT ( 9,12) OF BETA
```

McArdle and Aber (1990) demonstrate and discuss the relative fit for a family of growth curve models.

T-values and fitted moment and residuals for this model were

```
T-VALUES

        LAMBDA Y

              1.WISC_6   2.WISC_7   3.WISC_9   4.WISC11   5.ERR_6   6.ERR_7

WISC_6        0.000      0.000      0.000      0.000      0.000     0.000
WISC_7        0.000      0.000      0.000      0.000      0.000     0.000
WISC_9        0.000      0.000      0.000      0.000      0.000     0.000
WISC_11       0.000      0.000      0.000      0.000      0.000     0.000
  CONST.      0.000      0.000      0.000      0.000      0.000     0.000


        LAMBDA Y

              7.ERR_9    8.ERR11    9.LEVEL   10.SHAPE    CONST    12.STD.L

WISC_6        0.000      0.000      0.000      0.000      0.000     0.000
WISC_7        0.000      0.000      0.000      0.000      0.000     0.000
WISC_9        0.000      0.000      0.000      0.000      0.000     0.000
WISC_11       0.000      0.000      0.000      0.000      0.000     0.000
  CONST.      0.000      0.000      0.000      0.000      0.000     0.000


        LAMBDA Y

              13.STD.S

WISC_6        0.000
WISC_7        0.000
WISC_9        0.000
WISC_11       0.000
  CONST.      0.000


        BETA

              1.WISC_6   2.WISC_7   3.WISC_9   4.WISC11   5.ERR_6   6.ERR_7

1.WISC_6      0.000      0.000      0.000      0.000      0.000     0.000
2.WISC_7      0.000      0.000      0.000      0.000      0.000     0.000
3.WISC_9      0.000      0.000      0.000      0.000      0.000     0.000
4.WISC11      0.000      0.000      0.000      0.000      0.000     0.000
5.ERR_6       0.000      0.000      0.000      0.000      0.000     0.000
6.ERR_7       0.000      0.000      0.000      0.000      0.000     0.000
7.ERR_9       0.000      0.000      0.000      0.000      0.000     0.000
8.ERR11       0.000      0.000      0.000      0.000      0.000     0.000
9.LEVEL       0.000      0.000      0.000      0.000      0.000     0.000
10.SHAPE      0.000      0.000      0.000      0.000      0.000     0.000
CONST         0.000      0.000      0.000      0.000      0.000     0.000
12.STD.L      0.000      0.000      0.000      0.000      0.000     0.000
13.STD.S      0.000      0.000      0.000      0.000      0.000     0.000
```

BETA

	7.ERR_9	8.ERR11	9.LEVEL	10.SHAPE	CONST	12.STD.L
1.WISC_6	0.000	0.000	0.000	0.000	0.000	0.000
2.WISC_7	0.000	0.000	0.000	0.000	0.000	0.000
3.WISC_9	0.000	0.000	0.000	0.000	0.000	0.000
4.WISC11	0.000	0.000	0.000	0.000	0.000	0.000
5.ERR_6	0.000	0.000	0.000	0.000	0.000	0.000
6.ERR_7	0.000	0.000	0.000	0.000	0.000	0.000
7.ERR_9	0.000	0.000	0.000	0.000	0.000	0.000
8.ERR11	0.000	0.000	0.000	0.000	0.000	0.000
9.LEVEL	0.000	0.000	0.000	0.000	43.026	0.000
10.SHAPE	0.000	0.000	0.000	0.000	62.503	0.000
CONST	0.000	0.000	0.000	0.000	0.000	0.000
12.STD.L	0.000	0.000	0.000	0.000	0.000	0.000
13.STD.S	0.000	0.000	0.000	0.000	0.000	0.000

BETA

	13.STD.S
1.WISC_6	0.000
2.WISC_7	0.000
3.WISC_9	0.000
4.WISC11	0.000
5.ERR_6	0.000
6.ERR_7	0.000
7.ERR_9	0.000
8.ERR11	0.000
9.LEVEL	0.000
10.SHAPE	0.000
CONST	0.000
12.STD.L	0.000
13.STD.S	0.000

PSI

	1.WISC_6	2.WISC_7	3.WISC_9	4.WISC11	5.ERR_6	6.ERR_7
1.WISC_6	0.000					
2.WISC_7	0.000	0.000				
3.WISC_9	0.000	0.000	0.000			
4.WISC11	0.000	0.000	0.000	0.000		
5.ERR_6	0.000	0.000	0.000	0.000	14.283	
6.ERR_7	0.000	0.000	0.000	0.000	0.000	14.283
7.ERR_9	0.000	0.000	0.000	0.000	0.000	0.000
8.ERR11	0.000	0.000	0.000	0.000	0.000	0.000
9.LEVEL	0.000	0.000	0.000	0.000	0.000	0.000
10.SHAPE	0.000	0.000	0.000	0.000	0.000	0.000
CONST	0.000	0.000	0.000	0.000	0.000	0.000
12.STD.L	0.000	0.000	0.000	0.000	0.000	0.000
13.STD.S	0.000	0.000	0.000	0.000	0.000	0.000

PSI

	7.ERR_9	8.ERR11	9.LEVEL	10.SHAPE	CONST	12.STD.L
7.ERR_9	14.283					
8.ERR11	0.000	14.283				
9.LEVEL	0.000	0.000	0.000			
10.SHAPE	0.000	0.000	0.000	0.000		
CONST	0.000	0.000	0.000	0.000	0.000	
12.STD.L	0.000	0.000	0.000	0.000	0.000	8.486
13.STD.S	0.000	0.000	0.000	0.000	0.000	5.835

PSI

	13.STD.S
13.STD.S	5.177

```
FITTED MOMENTS AND RESIDUALS

        FITTED MOMENTS

              WISC_6      WISC_7      WISC_9      WISC_11     CONST.

WISC_6       414.877
WISC_7       506.075      646.218
WISC_9       721.745      908.475    1313.891
WISC_11      952.781     1201.065    1725.667    2298.520
  CONST.      19.224       24.367      35.234      46.875     1.000

        FITTED RESIDUALS

              WISC_6      WISC_7      WISC_9      WISC_11     CONST.

WISC_6       -44.528
WISC_7        -0.778       73.708
WISC_9       -43.850       50.037     -10.496
WISC_11      -59.281       64.778     -13.175     -20.272
  CONST.      -1.190        1.452       0.021      -0.282     0.000

        NORMALIZED RESIDUALS

              WISC_6      WISC_7      WISC_9      WISC_11     CONST.

WISC_6        -1.081
WISC_7        -0.015        1.149
WISC_9        -0.605        0.551      -0.080
WISC_11       -0.619        0.539      -0.077      -0.089
  CONST.      -0.606        0.587       0.006      -0.060     0.000
```

To exemplify choosing a different basis, we reran the model assuming equally spaced measures and using polynomial coefficients. The LISREL estimates for that model follow:

```
TITLE LINEAR LATENT GROWTH CURVE MODEL

LISREL ESTIMATES (MAXIMUM LIKELIHOOD)

        LAMBDA Y

              1.WISC_6    2.WISC_7    3.WISC_9    4.WISC11    5.ERR_6     6.ERR_7

WISC_6         1.000       0.000       0.000       0.000       0.000       0.000
WISC_7         0.000       1.000       0.000       0.000       0.000       0.000
WISC_9         0.000       0.000       1.000       0.000       0.000       0.000
WISC_11        0.000       0.000       0.000       1.000       0.000       0.000
  CONST.       0.000       0.000       0.000       0.000       0.000       0.000

        LAMBDA Y

              7.ERR_9     8.ERR11     9.LEVEL    10.SHAPE     CONST      12.STD.L

WISC_6         0.000       0.000       0.000       0.000       0.000       0.000
WISC_7         0.000       0.000       0.000       0.000       0.000       0.000
WISC_9         0.000       0.000       0.000       0.000       0.000       0.000
WISC_11        0.000       0.000       0.000       0.000       0.000       0.000
  CONST.       0.000       0.000       0.000       0.000       1.000       0.000
        LAMBDA Y
             13.STD.S

WISC_6         0.000
WISC_7         0.000
WISC_9         0.000
WISC_11        0.000
  CONST.       0.000
```

BETA

	1.WISC_6	2.WISC_7	3.WISC_9	4.WISC11	5.ERR_6	6.ERR_7
1.WISC_6	0.000	0.000	0.000	0.000	1.000	0.000
2.WISC_7	0.000	0.000	0.000	0.000	0.000	1.000
3.WISC_9	0.000	0.000	0.000	0.000	0.000	0.000
4.WISC11	0.000	0.000	0.000	0.000	0.000	0.000
5.ERR_6	0.000	0.000	0.000	0.000	0.000	0.000
6.ERR_7	0.000	0.000	0.000	0.000	0.000	0.000
7.ERR_9	0.000	0.000	0.000	0.000	0.000	0.000
8.ERR11	0.000	0.000	0.000	0.000	0.000	0.000
9.LEVEL	0.000	0.000	0.000	0.000	0.000	0.000
10.SHAPE	0.000	0.000	0.000	0.000	0.000	0.000
CONST	0.000	0.000	0.000	0.000	0.000	0.000
12.STD.L	0.000	0.000	0.000	0.000	0.000	0.000
13.STD.S	0.000	0.000	0.000	0.000	0.000	0.000

BETA

	7.ERR_9	8.ERR11	9.LEVEL	10.SHAPE	CONST	12.STD.L
1.WISC_6	0.000	0.000	1.000	-0.671	0.000	0.000
2.WISC_7	0.000	0.000	1.000	-0.224	0.000	0.000
3.WISC_9	1.000	0.000	1.000	0.224	0.000	0.000
4.WISC11	0.000	1.000	1.000	0.671	0.000	0.000
5.ERR_6	0.000	0.000	0.000	0.000	0.000	0.000
6.ERR_7	0.000	0.000	0.000	0.000	0.000	0.000
7.ERR_9	0.000	0.000	0.000	0.000	0.000	0.000
8.ERR11	0.000	0.000	0.000	0.000	0.000	0.000
9.LEVEL	0.000	0.000	0.000	0.000	31.424	1.000
10.SHAPE	0.000	0.000	0.000	0.000	21.260	0.000
CONST	0.000	0.000	0.000	0.000	0.000	0.000
12.STD.L	0.000	0.000	0.000	0.000	0.000	0.000
13.STD.S	0.000	0.000	0.000	0.000	0.000	0.000

BETA

	13.STD.S
1.WISC_6	0.000
2.WISC_7	0.000
3.WISC_9	0.000
4.WISC11	0.000
5.ERR_6	0.000
6.ERR_7	0.000
7.ERR_9	0.000
8.ERR11	0.000
9.LEVEL	0.000
10.SHAPE	1.000
CONST	0.000
12.STD.L	0.000
13.STD.S	0.000

PSI

	1.WISC_6	2.WISC_7	3.WISC_9	4.WISC11	5.ERR_6	6.ERR_7
1.WISC_6	0.000					
2.WISC_7	0.000	0.000				
3.WISC_9	0.000	0.000	0.000			
4.WISC11	0.000	0.000	0.000	0.000		
5.ERR_6	0.000	0.000	0.000	0.000	10.960	
6.ERR_7	0.000	0.000	0.000	0.000	0.000	10.960
7.ERR_9	0.000	0.000	0.000	0.000	0.000	0.000
8.ERR11	0.000	0.000	0.000	0.000	0.000	0.000
9.LEVEL	0.000	0.000	0.000	0.000	0.000	0.000
10.SHAPE	0.000	0.000	0.000	0.000	0.000	0.000
CONST	0.000	0.000	0.000	0.000	0.000	0.000
12.STD.L	0.000	0.000	0.000	0.000	0.000	0.000
13.STD.S	0.000	0.000	0.000	0.000	0.000	0.000

```
        PSI

             7.ERR_9    8.ERR11    9.LEVEL   10.SHAPE   CONST    12.STD.L
             ———————    ———————    ———————   ————————   —————    ————————
7.ERR_9       10.960
8.ERR11        0.000     10.960
9.LEVEL        0.000      0.000     0.000
10.SHAPE       0.000      0.000     0.000     0.000
CONST          0.000      0.000     0.000     0.000    1.000
12.STD.L       0.000      0.000     0.000     0.000    0.000     53.354
13.STD.S       0.000      0.000     0.000     0.000    0.000     19.892

        PSI

             13.STD.S
             ————————
13.STD.S      11.949
```

```
                MEASURES OF GOODNESS OF FIT FOR THE WHOLE MODEL :
      CHI-SQUARE WITH   9 DEGREES OF FREEDOM IS     90.48 (PROB. LEVEL = 0.000)
                      GOODNESS OF FIT INDEX IS 0.349
                 ADJUSTED GOODNESS OF FIT INDEX IS 0.623
                  ROOT MEAN SQUARE RESIDUAL IS    42.042
```

The new basis sits in **BETA**. The path from the constant to Level now holds the grand mean. The fit of the model is essentially the same (except for, we hope, some rounding error).

4. Conclusion

Estimating SEMs representing different models of change provides a powerful way to test complex hypotheses. The models, at first sight, may seem somewhat tricky even to those familiar with modeling techniques. The use of RAM notation, regardless of the algorithm selected, makes the logic of the different models somewhat easier to comprehend.

Chapter 5

Methods of Multidimensional Scaling

1. Summary of Method

To introduce methods of multidimensional scaling, Kruskal and Wish (1978) used the following analogy. Suppose you are given a map of the United States and you are asked to generate a table of distances between major cities. This is an easy task because you simply have to fill in the distances that can be measured with a ruler. Suppose, however, you are given this distance table and you are asked to generate a spatial representation of the location of these cities. This is a far more complicated task. Methods of multidimensional scaling perform this task

Scaling methods allow one to reduce proximities to the underlying dimensions. Proximities are measures of similarity or dissimilarity (distance). Typically, the number of dimensions is smaller than the number of objects. By "objects" we mean anything that can be measured, e.g., stimuli, cars, attitudes, concreteness, and belief systems.

Scaling methods calculate coordinates in the d-dimensional space from a proximity matrix. These coordinates are determined so that the relationship between each point of objects is as close as possible to the objects' similarity or dissimilarity as perceived by the subjects. Methods for the estimation of coordinates include least squares and other optimization algorithms.

Typically, researchers do not pursue the goal to perfectly match the empirical relationships between the objects. A perfect match is observed either if the number of dimensions is only two less then the number of objects or if fewer dimensions allow one to describe the pattern of relationships without loss of information. Only the second result satisfies the desideratum of scientific parsimony. Because of inconsistencies in subjects' judgments perfect matches rarely occur.

Scaling methods are called nonmetric if they only make a few assumptions of the data quality, e.g., in regard to numerical qualities. Specifically, most nonmetric scaling methods use the rank information of the similarity or dissimilarity judgments and transform this information into quantitative metric information. For most nonmetric methods there are metric alternatives (see Wood, 1990; Hartmann, 1979; MacCallum, 1988).

The distance between the calculated and the empirical relationships is measured by distance coefficients. The best known of these coefficients is Kruskal's (1964a, b) stress, s. The most common version of s can be developed as follows. It is the goal to estimate proximities δ_{ij} between objects i and j so that

$$f(\delta_{ij}) = d_{ij} \tag{1}$$

where d_{ij} denotes the perceived distance between objects i and j. The stress, s, indicates how good the mapping between $f(\delta_{ij})$ and d_{ij} is. The stress is

$$s = \sum_{i<j} (f(\delta_{ij}) - d_{ij})/t \tag{2}$$

where t denotes some scale factor; for instance, $t = (\sum_{i<j} d_{ij}^2)^{1/2}$. For other stress formulas see Wood (1990).

The magnitude of s is often evaluated as follows: $s = 0$ indicates perfect fit, $s \leq 0.025$ very good fit, $0.025 < s \leq 0.05$ good fit, $0.05 < s \leq 0.1$ relatively good fit, and $s > 0.1$ poor fit.

Multidimensional scaling (MDS) and principal component analysis are similar to each other in both function and application; however, they differ in important respects. From a user's perspective it is important that MDS makes fewer assumptions in respect to data quality and typically allows one to describe the pattern of relationships in fewer dimensions than principal component analysis.

The following sections discuss two applications of MDS. The first is a standard application of MDS to one individual's comparative judgments of dissimilarity of colors. SYSTAT (Wilkinson, 1988) is used for these analyses. The second is a simultaneous analysis of several subjects' judgments as discussed by Wood (1990). ALSCAL, a SAS procedure, is used for these analyses. There is a host of alternative multidimensional scaling procedures. Here we focus on methods for which programs are readily available. For more information see Borg (1981), Schiffman *et al.* (1981), or MacCallum (1988).

2. Computational Procedures for Standard Application of MDS

This section outlines the computational procedures for standard MDS using a microcomputer. In the next section, these procedures will be illustrated in detail using a data example. The software package SYSTAT will be used for both preparation of data and computation of the MDS solution.

2.1 COMPUTATIONAL PROCEDURES AND CONSIDERATIONS

MDS can be based on either direct or indirect comparisons of objects. People provide direct comparisons when they judge pairs of objects on a similarity or dissimilarity scale. Indirect comparisons are based on judgments of objects in regard to certain characteristics. These judgments can then be compared using similarity or dissimilarity coefficients (e.g., the product moment correlation or the Euclidean distance). All major software packages allow one to generate matrices containing coefficients of similarity or dissimilarity.

The measures of similariy or dissimilarity can be arranged in lower triangular matrices. The cells ij of these matrices contain the perceived or calculated proximity between objects i and j. The diagonals $i = j$ contain zeros for dissimilarity matrices and ones for similarity matrices. The upper triangle of these matrices is typically not considered because the measures are assumed to be symmetrical. In other words, researchers assume that the relationships between i and j are the same as the relationships between j and i.

There are only a few constraints on matrices for MDS, *e. g.*, that the matrix be Gramian, i.e., positive semi-definite. As a result, any coefficient of similarity or dissmilarity that is metric and provides a positive semi-definite matrix can be fed into MDS. Examples include sums of squares and cross-products matrices, covariance matrices, and matrices containing the following coefficients: Pearson correlation, gamma, Guttman µ2, Spearman rank, Kendall tau, normalized euclidean distances, positive matching dichotomy coefficient, Jaccard's dichotomy coefficient, simple matching dichotomy coefficient, Anderberg's dichotomy coefficient, and Tanimoto's dichotomy coefficient (Wilkinson, 1988).

A second important constraint that must be taken into account when preparing data for MDS concerns missing data. The pairwise deletion of cases with missing data can yield matrices with coefficients based on different subsamples. As a result, such matrices often are singular. Only listwise deletion of cases with missing data or proper imputation of missing values makes sure the coefficients are based on judgments from the same individuals.

The programs we discuss for MDS allow one to generate system files using spread sheet-like vignettes, to read in ASCII files written with word processors, or they provide routines for keying in data directly in system files. The following input information is needed for the MDS module: (1) tiangular matrix with coefficients of similarity or dissimilarity and (2) scaling method to perform. Options concerning the graphical output, the saving of output files, or the specification of start configurations are available.

For the examples in this chapter we use the SYSTAT package, release 4.0 (Wilkinson, 1988). SYSTAT 4.0 can be used with IBM and compatible microcomputers under DOS. The resident memory must have 512 K bytes in addition to the system.

2.2   AN ANNOTATED DATA EXAMPLE

This section analyzes a data example adopted from MacCallum (1988). A subject compared the ten colors red–purple (RP), red (R), yellow (Y), green–yellow–1 (GY1), green–yellow–2 (GY2), green (G), blue (B), blue–purple (PB), purple–1 (P1), and purple–2 (P2), with GY2 containing more green than GY1 and P2 more red than P1. The subject provided direct ratings of dissimilarity. The colors are approximately equally distributed around the color circle. We apply MDS to find out whether the comparative judgments of the subject reflect the color circle.

For the following steps we assume the dissimilarity ratings of the subject are available in form of a lower triangular matrix on the ASCII file COLOR.DAT. The statistical package SYSTAT will be used for analysis.

Step 1:   Preparation of Data

Generating a SYSTAT system file. To be able to apply statistical programs from the SYSTAT package, raw data must be transformed into a system file. Here we illustrate two of the options SYSTAT offers to generate a system file. The first is to read the data from an external ASCII file written with a word processor. (Another example is given in the chapter on prediction analysis; *cf.* Wilkinson, 1988.) Second, one can key in the data from within the DATA module.

Option 1: Reading Data from an External ASCII File. To generate an ASCII code file with a word processor, only a few rules must be observed. First, data points must be separated by blanks or commas, and second, the file must be saved in the ASCII code mode. Some word processors do not save files in ASCII code unless explicitly specified. The following example displays a print out of the file COLOR.DAT. The file contains the subject's judgments of dissimilarity. The diagonal cells contain zeros, thus indicating lack of dissimilarity of the colors with themselves. These zeros can be omitted for the following analyses.

```
0.000,        .   ,        .   ,        .   ,        .   ,
7.500,    0.000,        .   ,        .   ,        .   ,        .   ,
10.300,    6.900,    0.000,        .   ,        .   ,        .   ,
10.700,    8.500,    4.900,    0.000,        .   ,        .   ,
11.600,   10.700,    6.600,    3.500,    0.000,        .   ,
10.600,   11.100,    8.700,    6.300,    4.100,    0.000,
9.700,   12.200,   10.600,    7.800,    6.500,    5.000,
8.400,   10.800,   11.700,   10.400,    8.600,    7.400,
5.800,    9.900,   11.100,   11.600,   10.000,    9.100,
3.600,    8.000,   12.000,   11.300,   10.800,   10.700,
```

The following steps are performed within the DATA module, transform these data into a SYSTAT system file:

SAVE COLOR (Initiates the saving of the raw data system file called COLOR.SYS.)

INPUT RP,R,Y,GY1,GY2,G,B,PB,P1,P2 (Specifies names for variables.)

TYPE = DISSIMILARITY (Specifies the matrix as lower triangular containing dissimilarity judgments to be analyzed with the MDS module.)

GET COLOR (Reads the ASCII file COLOR.DAT.)

RUN (Saves the data as SYSTAT system file.)

If the zeros in the main diagonal are omitted, the command DIAGONAL ABSENT must be inserted after the TYPE specification.

Option 2: Keying in Data within the DATA Module. The DATA module allows one to key in data that then can be saved in system files or ASCII files. The commands for generating a system file are as follows:

SAVE COLOR (Initiates the saving of the raw data system file called COLOR.SYS.)

INPUT RP,R,Y,GY1,GY2,G,B,PB,P1,P2 (Specifies names for variables.)

TYPE = DISSIMILARITY (Specifies the matrix as lower triangular containing dissimilarity judgments to be analyzed with the MDS module.)

DIAGONAL ABSENT

RUN (Starts the process of keying in data. Program answers with the prompt to key in data row-wise. The prompt reads as follows: > Data input with return after each row. Data must be keyed in row-wise.)

NEW (Clears workspace for new operations and data. The command QUIT would do the same thing and then leave the module. SWITCHTO MDS transfers to the MDS module.)

The above commands generate a system file. The commands USE COLOR reads the system file COLOR.SYS in, and the command PUT COLOR generates the ASCII file COLOR.DAT. A third option for generating SYSTAT system files is to use the SYSTAT data editor. This editor provides a full screen vignette that allows one to key in the data directly. The system file COLOR.SYS generated through the first or the second option looks as follows:

		RP	R	Y	GY1	GY2
		G	B	PB	P1	P2
CASE	1	0.000	.	.	.	.
CASE	1	.	.	.	.	.
CASE	2	7.500	0.000	.	.	.
CASE	2	.	.	.	.	.
CASE	3	10.300	6.900	0.000	.	.
CASE	3	.	.	.	.	.
CASE	4	10.700	8.500	4.900	0.000	.
CASE	4	.	.	.	.	.
CASE	5	11.600	10.700	6.600	3.500	0.000
CASE	5	.	.	.	.	.
CASE	6	10.600	11.100	8.700	6.300	4.100
CASE	6	0.000	.	.	.	.
CASE	7	9.700	12.200	10.600	7.800	6.500
CASE	7	5.000	0.000	.	.	.
CASE	8	8.400	10.300	11.700	10.400	8.600
CASE	8	7.400	5.900	0.000	.	.
CASE	9	5.800	9.900	11.100	11.600	10.000
CASE	9	9.100	8.700	5.600	0.000	.
CASE	10	3.600	8.000	12.000	11.300	10.800
CASE	10	10.700	9.600	6.700	3.500	0.000

Step 2: Data Input

The MDS module in SYSTAT expects a lower triangular matrix as input. The matrix must be in a system file generated, for example, as shown above. The matrix must contain coefficients of the type discussed above.

Step 3:   Running the Program

Before running the program, the user must make a series of decisions. These decisions concern the number of dimensions, the type of scaling, the variables to be included, parameters that control the iterations, the saving of result files and the graphical presentation of results. The number of dimensions is comparable to the number of factors in factor analysis. Default value is 2. The program can handle up to 15 dimensions. As types of scaling, Kruskal's stress formula 1 and Goodman's coefficient of alienation are available. The latter method seems to be less susceptible to local minima. Variables can be selected by providing a variable list. Omitting this list leads to the inclusion of all variables. The parameters the user can fix to control the iteration include the maximum number of iterations, the minimum parameter value, and the minimum size of the decrement required for continuation of the search for a minimum. Results that can be saved on files include the final configuration, the residuals, and the estimated distances. In regard to the graphical output, one has the choice between plotting the distances against the estimated distances (option SHEPARD = 1) and plotting the distances against the estimated distances, separately for each point of the pair (option SHEPARD = 2). The program allows the user to input initial configurations. This option is useful, for instance, to avoid degenerate solutions or to perform confirmatory analysis for which one lets the program iterate only once.

A sample printout for the color data is given below. It was generated using the default values for all parameters. The commands for this run are

USE COLOR (Reads the distance matrix from file COLOR.SYS.)

OUTPUT @ (Sends output to both printer and screen.)

SCALE (Starts MDS of all variables; Kruskal's stress is applied.)

The first part of the printout summarizes the iterations that led to a minimum of the stress. The output, below, shows that the program needed 19 iterations (default maximum is 50). After about the tenth iteration, the value for the stress seemed to stabilize. If there are no longer changes in the stress that are greater than the minimum decrement, the iteration will cease. This was obviously the case for $s = 0.02006$. A minumum value for the stress itself was not specified. Therefore, the program would have iterated until the maximum number of iterations had been reached, the stress had reached a stable value, or the stress had assumed the value zero. According to the above criteria, the result for the present data shows a very good fit.

```
MONOTONIC MULTIDIMENSIONAL SCALING

MINIMIZING KRUSKAL STRESS (FORM 1) IN 2 DIMENSIONS

ITERATION   STRESS
---------   ------
    0        .035
    1        .029
    2        .025
    3        .023
    4        .022
    5        .022
    6        .021
    7        .021
    8        .021
    9        .021
   10        .020
   11        .020
   12        .020
   13        .020
   14        .020
   15        .020
   16        .020
   17        .020
   18        .020
   19        .020

STRESS OF FINAL CONFIGURATION IS:        .02006
```

The next part of the output contains the so-called Shepard diagram (option SHEPARD = 1). It regresses the calculated to the empirically observed dissimilarities. For the present data we notice that the regression line is almost perfectly linear. In other words, large calculated dissimilarities correspond with large observed dissimilarities. This is the desired pattern.

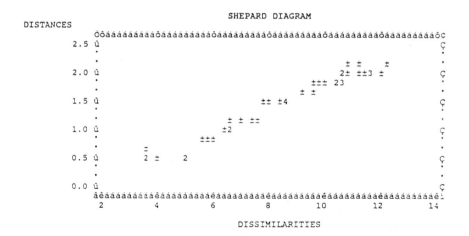

The third part of the print out gives the coordinates of the colors in the two-dimensional space that results from MDS. The program assignes to each of the variables a letter that is used for the graphical representation of the solution.

```
COORDINATES IN 2 DIMENSIONS

VARIABLE    PLOT    DIMENSION
--------    ----    ---------
                      1      2
   RP        A      -.92   -.45
    R        B      -.18  -1.20
    Y        C       .79   -.78
  GY1        D       .93   -.30
  GY2        E       .89    .21
    G        F       .62    .59
    B        G       .27    .90
   PB        H      -.42    .80
   P1        I      -.93    .29
   P2        J     -1.05   -.06
```

The two-dimensional graphical representation of the MDS solution shows that the subject's dissimilarity judgments nicely represent the color circle. This result is hard to realize by just inspecting the calculated coordinates, above (cf. MacCallum, 1988, p. 423).

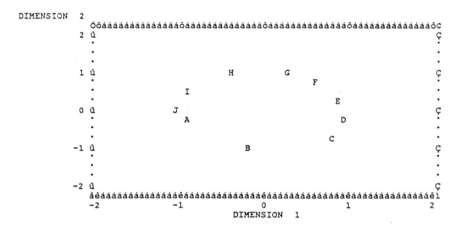

3. Computational Procedures for Three-Dimensional Analysis

Wood (1990; *cf.* Arabie *et al.* 1987) discusses methods that allow one to analyze similarity or dissimilarity data provided by several subjects. The MDS program discussed in the last section allows one to analyze group data only if they are aggregated such that only one similarity matrix results. The advantage of the three-dimensional methods is that matrices from several subjects can be considered separately, a common spatial representation can be calculated, and, in addition, the relative position of each subject can be determined. The ALSCAL program (Takane *et al.* 1976) in the SAS software package will used for the following sample analyses.

3.1 Outline of Computational Procedures

ALSCAL uses lower triangular matrices as input. The matrices are of the same form as the matrices used by SYSTAT. Matrices for several subjects are appended to each other. Similarity or dissimilarity matrices can be analyzed.

The ALSCAL program we discuss here is implemented on main-frame computers. It reads data from SAS system files. The following input information is needed: (1) triangular matrix or matrices containing (dis) similarity coefficients and (2) indication of scaling method to perform. Options concerning the graphical output, saving of output files, or algorithmic routines are available. ALSCAL is a SAS procedure. Thus, the usual SAS routines apply.

3.2 An Annotated Data Example

The following example was provided by Phillip K. Wood (Dept. of Psychology, University of Missouri, Columbia). Details concerning the SAS specifications, the program setup, and the SAS source code for the regression approaches presented in Wood (1990) can be requested from him or the author. The data used here are the same as in Wood (1990). Four children were presented with with nine glasses of water. These glasses differed in height and diameter, but they all contained the same amount of water. The children compared the glasses pairwise and rated by how much the water content differed between the two glasses in each pair.

Step 1: Data Preparation

The data provided by the four children are direct dissimilarity ratings. These data are conveyed to the ALSCAL program within SAS using the DATA statement. The use of the statement is illustrated in the sample printout below. It contains specifications of the file name (DISSIMS), the variable names (STIMID through $S31$, where STIMID is the name for the stimuli, or the stimulus identifier, and $S13$ through $S31$ identifies the comparison stimulus), the statement CARDS, followed by the raw data, and, after the data, the command to print the data. The following commands prepare the ALSCAL run:

```
DATA DISSIMS;
INPUT STIMID S13 S23 S33 S12 S22 S32 S11 S21 S31;
CARDS;
13 . . . . . . . . .
23 3 . . . . . . . .
```

```
33  5  3  .  .  .  .  .  .  .
12  1  4  5  .  .  .  .  .  .
22  1  2  5  2  .  .  .  .  .
32  3  0  3  4  2  .  .  .  .
11  2  5  5  1  3  5  .  .  .
21  1  4  5  0  2  4  1  .  .
31  0  3  5  1  1  3  2  1  .
13  .  .  .  .  .  .  .  .  .
23  0  .  .  .  .  .  .  .  .
33  0  0  .  .  .  .  .  .  .
12  3  3  3  .  .  .  .  .  .
22  3  3  3  0  .  .  .  .  .
32  3  3  3  0  0  .  .  .  .
11  5  5  5  3  3  3  .  .  .
21  5  5  5  3  3  3  0  .  .
31  5  5  5  3  3  3  0  0  .
13  .  .  .  .  .  .  .  .  .
23  3  .  .  .  .  .  .  .  .
33  5  3  .  .  .  .  .  .  .
12  0  3  5  .  .  .  .  .  .
22  3  0  3  3  .  .  .  .  .
32  5  3  0  5  3  .  .  .  .
11  0  3  5  0  3  5  .  .  .
21  3  0  3  3  0  3  3  .  .
31  5  3  0  5  3  0  5  3  .
13  .  .  .  .  .  .  .  .  .
23  3  .  .  .  .  .  .  .  .
33  3  3  .  .  .  .  .  .  .
12  1  4  5  .  .  .  .  .  .
22  1  2  5  2  .  .  .  .  .
32  3  0  3  3  1  .  .  .  .
11  2  3  5  1  3  5  .  .  .
21  1  3  5  1  1  1  .  .  .
31  1  2  3  1  1  1  2  1  .
PROC PRINT;
TITLE 'LISTING OF STIMULI DISSIMILARITIES FOR FOUR SUBJECTS';
DATA DISSIMS; SET DISSIMS; DROP STIMID;
PROC ALSCAL MODEL=INDSCAL DIMENS=2 PRINT PLOT HEADER;
```

Step 2: Data Input

ALSCAL can process lower triangular matrices for one or several individuals. The data statement in the present example was followed by the matrices for four children. The following sample printout was generated by the following statement:

PROC ALSCAL MODEL = INDSCAL DIMENS = 2 PRINT PLOT HEADER;

This statement starts a run of the ALSCAL procedure. Specifically, the INDSCAL model is used (Carroll and Chang, 1970). The DIMENS=2 statement specifies that we calculate a two-dimensional representation of the children's judgment, assuming the children use height and diameter as their chief criteria. PRINT sends the output to the printer, PLOT has the program generate pictorial representations which are also sent to the printer, and HEADER prints the title specified at the end of the last sample printout on the top of each page. The resulting printout contains the following information:

    1. the raw data as given in the DATA statement,

2. a summary of the program options,
3. the raw data separately for each subject,
4. the history of the iteration process including the resulting total stress and the stress for each subject,
5. the stimulus coordinates in two dimensions,
6. the so-called subjects' weights—that is, the weight each dimension contributes to each subject's judgments,
7. a picture of the two-dimensional representations of the stimuli as perceived by the subjects,
8. a picture of the dissimilarities between the glasses, subjectwise,
9. the scaled dissimilarities separately for each subject
10. the plot of the calculated and observed dissimilarities,
11. the one-dimensional subject weights, and
12. a plot of the one-dimensional subject weights.

From this detailed output, we include items 2, 5–8, and 10. The summary of the program options appears below.

```
LISTING OF STIMULI DISSIMILARITIES FOR FOUR SUBJECTS

11:45 MONDAY, JANUARY 22, 1990     2

PROC ALSCAL JOB OPTION HEADER


DATA OPTIONS-

NUMBER OF ROWS (OBSERVATIONS/MATRIX).      9
NUMBER OF COLUMNS (VARIABLES) .  .  .      9
NUMBER OF MATRICES    .  .  .  .  .  .     4
MEASUREMENT LEVEL .  .  .  .  .  .  .      ORDINAL
DATA MATRIX SHAPE .  .  .  .  .  .  .      SYMMETRIC
TYPE  .  .  .  .  .  .  .  .  .  .  .      DISSIMILARITY
APPROACH TO TIES  .  .  .  .  .  .  .      LEAVE TIED
CONDITIONALITY .  .  .  .  .  .  .  .      MATRIX
DATA CUTOFF AT  .  .  .  .  .  .  .  .     0.0


MODEL OPTIONS-

MODEL .  .  .  .  .  .  .  .  .  .  .      INDSCAL
MAXIMUM DIMENSIONALITY .  .  .  .  .       2
MINIMUM DIMENSIONALITY .  .  .  .  .       2
NEGATIVE WEIGHTS .  .  .  .  .  .  .       NOT PERMITTED


OUTPUT OPTIONS-

JOB OPTION HEADER .  .  .  .  .  .  .      PRINTED
DATA MATRICES  .  .  .  .  .  .  .  .      PRINTED
CONFIGURATIONS AND TRANSFORMATIONS  .     PLOTTED
OUTPUT DATASET .  .  .  .  .  .  .  .      NOT CREATED
INITIAL STIMULUS COORDINATES  .  .  .     COMPUTED
INITIAL COLUMN STIMULUS COORDINATES .     COMPUTED
INITIAL SUBJECT WEIGHTS .  .  .  .  .      COMPUTED
INITIAL STIMULUS WEIGHTS  .  .  .  .       COMPUTED


ALGORITHMIC OPTIONS-

MAXIMUM ITERATIONS  .  .  .  .  .  .       30
CONVERGENCE CRITERION  .  .  .  .  .       0.00100
MINIMUM S-STRESS .  .  .  .  .  .  .       0.00500
MISSING DATA ESTIMATED BY .  .  .  .       ULBOUNDS
TIESTORE .  .  .  .  .  .  .  .  .  .      144
```

The sample listing shows that the user must make a number of decisions concerning the scaling model, output, and the algorithms used. As is obvious from the above PROC statement, we used chiefly the default values for our example. Specifically, we used the INDSCAL model, requested only one, a two-dimensional solution, and did not permit negative weights. In the output options we request that the job option header specified earlier be printed; data matrices are also printed, and configurations and transformations are depicted. An output file is not created. The program calculates start values for stimulus coordinates and subject and stimulus weights. (Another option would be to determine the start values from substantive theory.)

In the algorithmic options section we see that the default value for the maximum number of iteration steps is 30. For complex solutions and large data sets this number may be increased. The user may also specify the convergence criterion, that is, the smallest difference between two solutions that must be exceeded for continuation of the iterations, and the minimum stress that is the stress value that terminates the iterations.

The iteration history for the present example (not listed here) shows that the program needed only five iterations. After the fourth iteration the stress improvement was less than 0.001. Therefore, the program terminated the iteration process. If we assume that the solution is not a local minimum, we may not expect substantial improvements from continuing the iteration process. The stresses for the four children range between 0.070 and 0.278 with an average of 0.202, indicating relatively good fit for only one child. To improve the fit we could have increased the number of dimensions for the calculated solution.

The following part of the output gives the two-dimensional stimulus coordinates for the nine glasses as perceived by all children together.

```
LISTING OF STIMULI DISSIMILARITIES FOR FOUR SUBJECTS

11:45 MONDAY, JANUARY 22, 1990    8

CONFIGURATION DERIVED IN 2 DIMENSIONS


        STIMULUS COORDINATES

                         DIMENSION
STIMULUS   PLOT        1          2
NUMBER    SYMBOL

    1        1       -1.0460     1.1256
    2        2       -1.2484    -0.0891
    3        3       -1.4205    -1.3301
    4        4        0.1056     1.2669
    5        5        0.1778     0.0696
    6        6       -0.1440    -1.2880
    7        7        1.1902     1.2356
    8        8        1.2418     0.0633
    9        9        1.1435    -1.0739
```

Again, a look at the printed coordinates is less informative than a look at the picture of the location of the objects in the two-dimensional space. Insert A depicts the ALSCAL solution.

The picture indicates that, overall, the children reproduced the two-dimensional pattern very well. Thus, we tend to use this solution even though the overall stress suggests a rather poor fit.

On major source for poor fits in multiple subjects-MDS lies in the differences between subjects' judgments. Differences between subjects are indicated, for instance, by differences in the weight each dimension has for each subject. These weights are given in Insert B.

The weights printed for each subject suggest that for the first and the last child each dimension contributes about equally. Thus, we may conclude that, for these two children, height and diameter are of about equal importance to their dissimilarity ratings. In contrast to these two children, the other two rely predominantly on only one dimension. Child number 2 uses mainly height, and child number 3 mainly uses diameter information for his/her ratings. These interindividual differences explain a large portion of the overall large stress. (Notice that the header for the table in the printout is a misnomer. The table contains the weights that the dimensions carry, but not the subjects' weights.) The weirdness values indicate how inbalanced the subjects' ratings are, relative to the average weights. Large values indicate that only one dimension has a large weight. Small values indicate that the dimensions' weights are proportional to the average weights.

The next figure in the printout (Insert C) depicts the subjects' weights. Notice that the scaling of the dimensions is not perfect. The third child's values do not fit in the frame. Therefore, the number 3 does not appear in the picture.

The next printout item reproduced (Insert D) contains the scatterplot of the calculated values by the observed values or, in ALSCAL terms, the disparities by the distances scatterplot. The picture shows the data for all children.

Overall, the scatterplot suggests that the calculated solution has some of the desired characteristics. Large distances tend to go hand in hand with large disparities; however, as the high stress indicates, this solution only provides a poor fit. The scatterplot does not indicate what points were generated by what child. The calculated disparity matrices which are not reproduced here show that the disparities of the first child vary between 1.220 and 1.939. The ranges for the other children are 0.316 -2.415 for child 2, 0.294 - 2.437 for child 3, and 1.316 - 1.887 for child 4.

Other programs for scaling include the Guttman–Lingoes nonmetric program series (Lingoes, 1973). The SAS source code for the regression-type analyses performed by Wood (1990) can be requested from Dr. Wood.

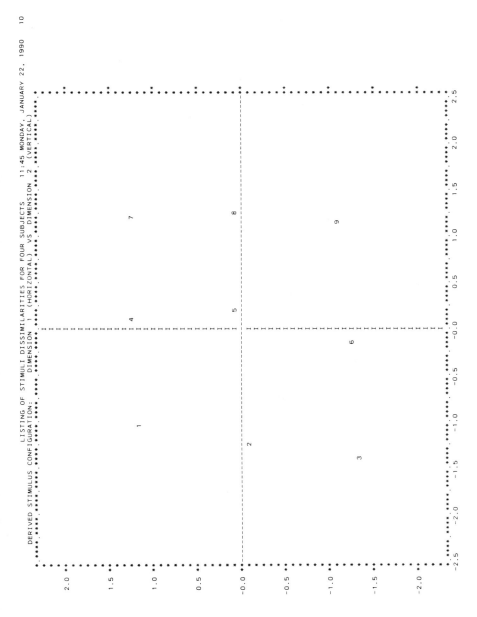

INSERT A

LISTING OF STIMULI DISSIMILARITIES FOR FOUR SUBJECTS   11:45 MONDAY, JANUARY 22, 1990   9

SUBJECT WEIGHTS MEASURE THE IMPORTANCE OF EACH DIMENSION TO EACH SUBJECT. SQUARED WEIGHTS SUM TO RSQ.

A SUBJECT WITH WEIGHTS PROPORTIONAL TO THE AVERAGE WEIGHTS HAS A WEIRDNESS OF ZERO, THE MINIMUM VALUE.
A SUBJECT WITH ONE LARGE WEIGHT AND MANY LOW WEIGHTS HAS A WEIRDNESS NEAR ONE.
A SUBJECT WITH EXACTLY ONE POSITIVE WEIGHT HAS A WIERDNESS OF ONE, THE MAXIMUM VALUE FOR NONNEGATIVE WEIGHTS.

SUBJECT WEIGHTS

SUBJECT NUMBER	PLOT SYMBOL	WEIRD-NESS	DIMENSION 1	DIMENSION 2
1	1	0.0116	0.4731	0.4553
2	2	0.9746	0.9769	0.0191
3	3	0.9727	0.0216	0.9880
4	4	0.0308	0.4333	0.4046

OVERALL IMPORTANCE OF EACH DIMENSION   0.3416   0.3369

INSERT B

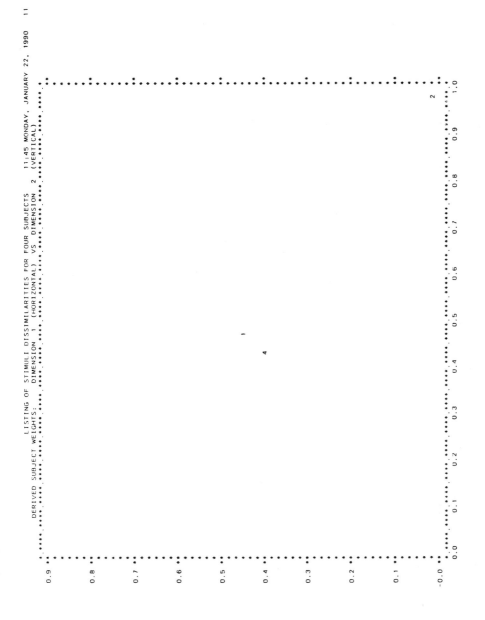

INSERT C

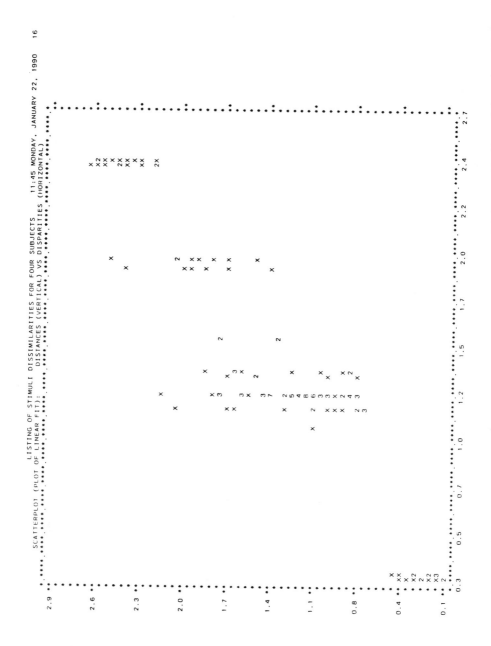

INSERT D

Chapter 6

Event-History Analysis (Example: Survival Analysis)

1. Summary of Method

In many applications of social science statistics, researchers are interested in the amount of time it takes before a dependent variable assumes another state. For instance, we might ask how long a patient is willing to sit in waiting rooms before he/she gets upset and leaves, how long a new car runs flawlessly, how old people are when they marry for the first time, or how many miles we can drive before we get a flat tire. This type of event can be analyzed using event history analysis (Allison, 1984; Blossfeld *et al.* 1988; Petersen, 1990; Tuma and Hannan, 1984).

The following three characteristics determine the type of event history analysis performed: type of event, type of change, and concept of time in data collection. Event-history analysis deals with events that either are single and nonrepeatable or can occur repeatedly. Most prominent among the former are deaths. Deaths are of major interest in medicine, biology, and sociology. Examples of repeatable events include job changes and marriages.

The second issue of concern is the number of states a variable can possibly assume after a change. For instance, quitting a job can result in being jobless, self-employed, employed, or retired. The third issue concerns the concept of time in data collection and analysis. When time is measured in

105

large units, one estimates discrete time models. When time is measured in very small units, one estimates a continuous time model.

Statistical issues in event-history analysis involve the assumptions one is willing to make when estimating parameters. Virtually no assumptions need to be made in the application of nonparametric models. For semiparametric models one makes only a few assumptions. For instance, for Cox regression models one uses regression models that have a specific form. However, the type of event-time distribution is not specified. For parametric models one usually assumes that the time variable follows a specific distribution, e.g., the exponential, Weibull, log-normal, or log-logistic distributions.

Life-event data, e.g., job changes, are often sampled in retrospect. As a result, the state change may not have occurred at the time of interview. For instance, the individual may not have married, or the car stills runs flawlessly. Data of this type have been termed "right-censored." Another type of data has been termed "left-censored." This type of data describes life events from a certain point in time on. The problem here is that it may be unknown how much time an individual has spent in a state before the beginning of the observations. For instance, it may be unknown when an individual was infected with a virus.

Event-history analysis comprises a wide range of statistical methods. (See Blossfeld *et al.*, 1988; Petersen, 1990.) To keep the present discussion simple, we focus on methods for survival analysis, that is, for single, nonrepeatable events. Also, we confine ourselves to right-censored data. Survival analysis estimates the probability for a case not to have changed state after having been observed m times. Between the observations, we have the time intervals

$$[0, t_1], [t_1, t_2], ..., [t_{m-1}, t_m]. \qquad (1)$$

The probability of not having changed state at point in time j is

$$p_j = 1 - 1/r_j, \qquad (2)$$

where r_j denotes the number of cases that did not change state until time j. The Kaplan–Meier estimator of the survivor function is

$$S(t_i) = \begin{cases} 1 & \text{for } t_j < t_i \\ \prod_{t_j < t_i} p_j & \text{otherwise.} \end{cases} \qquad (3)$$

The product in (2) applies to the set of observation points t for which $t_j < t_i$.

If more than one event can occur at a time, we substitute

$$p_j = 1 - d_j / r_j \qquad (4)$$

for (2), where d_j denotes the number of events possible at time j. The asymptotic variance of the Kaplan-Meier estimator, necessary for statistical inference, is

$$s_2 = S(t_i)^2 \sum_{t_j < t_j} \frac{d_j}{r_j(r_j - d_j)} \ . \qquad (5)$$

The expectancy of the estimator is

$$\mu = \sum_j S(t_{i-1})(t_i - t_{i-1}) \qquad (6)$$

where the summation goes over all observations. The observations must be rank ordered. This estimator tends to overestimate the true expectancy if t_i is censored.

Generalizations of the estimator have been developed by Turnbull (1976). The Kaplan-Meier estimator is nonparametric. Semiparametric approaches to event history analysis include regression models, e.g., Cox's (1972) regression models with covariates. Parametric models allow one to fit fully specified probability models that contain a finite number of unknown parameters (See Petersen, 1990; Blossfeld et al., 1988). For discussions from a reliability point of view, see, e.g., Miller (1981).

2.   Outline of the Computational Procedures

This section outlines the steps for computing survival analysis using a PC. The next section illustrates these steps in detail using a data example. The program SURVIVAL by Steinberg and Colla (1988) in the version distributed by SYSTAT will be used.

The program SURVIVAL is distributed as a supplement to SYSTAT. As such, it reads data only from SYSTAT system files which can be generated using the SYSTAT DATA or EDIT modules. To use SURVIVAL, the user needs a file that contains only two variables. The first one is the dependent variable. This variable measures, for instance, the length of survival of a patient with terminal illness or miles flawlessly driven by a car. The dependent variable must be positive. The second variable functions as coding variable for the event under study. It assumes a 1 as long as the critical event has not occurred. It assumes a 0 when the event has occurred. The coding variable is also called censoring variable because it indicates the time period censored by the occurrence of the critical event. Additional variables that can be analyzed by SURVIVAL include stratification variables and covariates.

All SYSTAT modules read data from system files. The command "USE filename" starts the data input. Specifications that control a particular run are inputted interactively or read from an ASCII code file with the suffix .CMD. This file contains the commands exactly as the user would input them via the keyboard. Each command begins at a new line. Commands that occupy more than one line are split by a comma.

After data input the user can select from several options for survival analysis. Stratification of subject samples and the analysis of covariates are among these options. The annotated data example in the next section will explore some options in detail. Minimally, one must specify which variable is the dependent variable and which is the coding variable. These specifications use the following commands:

TIME = variable name (Specifies the dependent variable.)

CENSOR = variable name (Specifies the coding variable.)

The Kaplan-Meier estimator of the survivor curve is calculated after issuing KM. This command will result in a table and a plot of the data. The resulting table is a lifetable.

SURVIVAL offers a plethora of options. For instance, it can handle right-censored, interval-censored and right-censored, and other type data. The program analyzes data only if they are sorted. The primary sort key is the TIME variable, the secondary key is the censoring variable, and the last key is the LTIME variable (if present). The TIME variable specifies the upper bound of the censoring interval; it is sorted in ascending order. The LTIME variable specifies the lower bound of the censoring interval. It is sorted in descending order. The censoring variable is sorted in the order 1, -1, 0. SURVIVAL will sort variables. However, large data sets should be sorted using the fast sorting module SSORT in SYSTAT.

The program offers nonparametric, semiparametric, and fully parametric analysis options. The nonparametric options include the Kaplan-Meier estimator, plots, stratification, rank tests, subset selection, histograms of the exact failure and censoring pattern, and Turnbull estimation. Semiparametric analysis options include Cox regressions (only for exact failure data or right-censored data), stratification, and plots. Fully parametric analysis options allow one to specify probability models which are fitted using maximum likelihood methods. The available models are based on the exponential, Weibull, log-normal, and log-logistic distributions. In all models one can include covariates. Model selection can be performed in a stepwise fashion with both forward and backward covariate selection strategies available. Also, covariates can be forced into the model. Point estimation of hazards is possible.

3. An Annotated Data Example

The following artificial data example illustrates only a few options of the SURVIVAL program. The data describe four brands of cars. The observed variables are TIME (number of months a car was used before being wrecked), CENSOR (a 1 indicates the car was still operational when observed the last time), MILES (number of miles driven at last observation), and MAKE (four brands). The data will be used to estimate the survivor function which indicates how likely it is for a car to still be functional after a given time. The number of miles driven will be used as a covariate, the brand as a stratification variable.

Steps 1 and 2:   Data Preparation and Input

The data for the present example were keyed in using the DATA module in SYSTAT. The following commands were used:

SAVE CAR (Initiates saving of data in system file CAR.SYS.)

INPUT TIME, CENSOR, MILES, MAKE (Names the variables in the file.)

RUN (Starts data input. Program prompts for data. Data can be keyed in with blanks or commas between. Data are keyed in casewise. Commas terminate a line if more than 80 columns per case are needed.)

NEW (Terminates data input. Data are saved. Workspace is cleared.)

The resulting SYSTAT system file appears below. Notice that this data file is unsorted.

		TIME	CENSOR	MILES	MAKE
CASE	1	80.000	1.000	85.000	1.000
CASE	2	90.000	1.000	92.000	2.000
CASE	3	121.000	1.000	111.000	4.000
CASE	4	60.000	1.000	54.000	3.000
CASE	5	65.000	1.000	77.000	3.000
CASE	6	99.000	1.000	123.000	4.000
CASE	7	76.000	1.000	76.000	2.000

CASE	8	69.000	1.000	81.000	4.000
CASE	9	87.000	1.000	87.000	1.000
CASE	10	90.000	1.000	66.000	1.000
CASE	11	102.000	1.000	102.000	4.000
CASE	12	54.000	1.000	79.000	3.000
CASE	13	80.000	1.000	99.000	2.000
CASE	14	130.000	1.000	180.000	4.000
CASE	15	101.000	1.000	120.000	1.000
CASE	16	109.000	1.000	133.000	2.000
CASE	17	96.000	1.000	110.000	3.000
CASE	18	89.000	1.000	95.000	2.000
CASE	19	76.000	1.000	90.000	1.000
CASE	20	88.000	1.000	114.000	3.000
CASE	21	110.000	1.000	76.000	4.000
CASE	22	66.000	1.000	73.000	2.000
CASE	23	96.000	1.000	118.000	3.000
CASE	24	69.000	1.000	120.000	1.000
CASE	25	73.000	1.000	53.000	3.000
CASE	26	89.000	1.000	141.000	4.000
CASE	27	34.000	1.000	44.000	3.000
CASE	28	99.000	1.000	125.000	4.000
CASE	29	87.000	1.000	89.000	1.000
CASE	30	90.000	1.000	115.000	1.000
CASE	31	88.000	1.000	106.000	2.000
CASE	32	101.000	1.000	121.000	2.000

Step 3: Running the Program

The following commands tell the program to read the data in. This step is necessary because SURVIVAL is not part of the standard SYSTAT package. Therefore, the SWITCHTO command cannot be used.

OUTPUT @ (Sends output to printer.)

USE CAR (Reads data from system file CAR.SYS. Program displays variables available to user.)

The data must be sorted before analysis. The program gives a message if it notices that the data are unsorted. Since the present data set is small, we use SURVIVAL for sorting and issue the following commands:

SORT=YES (Starts sorting of data. Sorting keys are as specified above.)

INPUT* (The asterisk indicates that all variables are used in subsequent analyses. The first group contains the special variables, that is, the variables used as sorting keys. All other variables are considered covariates. All variables are loaded into resident memory. If variables are listed after INPUT, only these will be loaded.)

The program responds by giving a summary of the data loaded into resident memory. This summary appears below.

```
RESETTING ALL OPTIONS.

CLEARING WORKSPACE, PLEASE WAIT.

CLEARING WORKSPACE, PLEASE WAIT.

WEIGHT DEFAULTS TO 1.0 FOR EACH OBSERVATION.
```

```
LOADING DATA

COVARIATES INPUT:
===========================================================================
     MILES          MAKE

SPECIAL VARIABLES INPUT:
===========================================================================
   TIME  =      TIME
 CENSOR  =    CENSOR
 WEIGHT  = 1.0

      32 INPUT OBSERVATIONS PROCESSED.

                                        WEIGHTED
 CENSORING            OBSERVATIONS    OBSERVATIONS
 ================================================
 EXACT FAILURES    |      32.000  |      32.000
 RIGHT CENSORED    |       0.000  |       0.000
 ================================================
          TOTALS   |      32.000  |      32.000

 SORTING DATA DURING INPUT.
 COVARIATE MEANS
 ================================================================
    MILES    =      98.594   |     MAKE    =       2.500

 TYPE 1, EXACT FAILURES AND RIGHT CENSORING ONLY.
 ANALYSES/ESTIMATES: KAPLAN-MEIER, COX AND PARAMETRIC MODELS

 OVERALL TIME RANGE: [ 34.000 , 130.000 ]
 FAILURE TIME RANGE: [ 34.000 , 130.000 ]
```

The first statement that concerns the present data specifies that each case gets the same weight. This results from not using the WEIGHT command which specifies which variable contains the case weights. The WEIGHT command specifies a WEIGHT variable which typically contains case frequencies, although it is not restricted to natural numbers.

Next, the program lists the covariates and special variables and gives the number of observations read in. A table follows specifying the censoring pattern of the data. The covariate means show that the average car in the sample had driven 98,594 miles. The means for MAKE are meaningless because MAKE is a nominal-level variable, specifying the car brand. The program indicates after the means what analyses can be performed with the inputted data. The time-range interval indicates that the cars under study were between 34 and 130 months old.

The next step after reading data is the estimation of the survivor function using, for instance, the Kaplan–Meier estimator. The command KM yields the following printout.

```
KAPLAN-MEIER ESTIMATION
ALL THE DATA WILL BE USED

       NUMBER        NUMBER                    K-M           STANDARD
       AT RISK       FAILING        TIME    PROBABILITY       ERROR
================================================================================
       32.000        1.000        34.000      0.969          0.031
       31.000        1.000        54.000      0.937          0.043
       30.000        1.000        60.000      0.906          0.052
       29.000        1.000        65.000      0.875          0.058
       28.000        1.000        66.000      0.844          0.064
       27.000        2.000        69.000      0.781          0.073
       25.000        1.000        73.000      0.750          0.077
       24.000        2.000        76.000      0.687          0.082
       22.000        2.000        80.000      0.625          0.086
       20.000        2.000        87.000      0.562          0.088
       18.000        2.000        88.000      0.500          0.088
       16.000        2.000        89.000      0.438          0.088
       14.000        3.000        90.000      0.344          0.084
       11.000        2.000        96.000      0.281          0.079
        9.000        2.000        99.000      0.219          0.073
        7.000        2.000       101.000      0.156          0.064
        5.000        1.000       102.000      0.125          0.058
        4.000        1.000       109.000      0.094          0.052
        3.000        1.000       110.000      0.063          0.043
        2.000        1.000       121.000      0.031          0.031
        1.000        1.000       130.000      0.000

GROUP SIZE = 32.000
NUMBER FAILING = 32.000
PRODUCT LIMIT LIKELIHOOD = -95.131

MEAN SURVIVAL TIME = 86.375

FAILURE QUANTILES
------------------
75%        99.000
50%        89.000
25%        76.000


KAPLAN-MEIER ESTIMATION
```

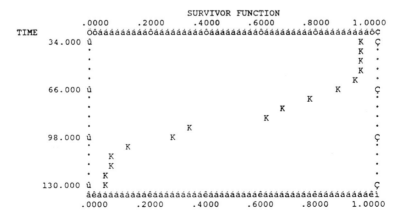

```
PLOT TYPE:    CUMULATIVE HAZARD
HPLOT PARAMETERS:             TMIN =        34.000
      DELT =        0.000   TMAX =       130.000
      PMIN =        0.005   PMAX =         5.298
      NLINES =     15        ADJUST
```

The table in this printout lists five variables. The first, "number at risk," tells us how many cars are still going strong at a given point in TIME. At the first point in time, that is, after 34 months, all but one car are still there. Thus, in the second time interval, the number of cars on the road, or at risk, is reduced by 1. The second interval covers the span from 34 to 54 months.

The Kaplan–Meier estimator suggests that, at the first interval, the probability for a car to be intact is 0.969. The standard error is 0.031. The Kaplan–Meier probabilities describe a step function. It jumps at each exact failure time. Toward the end of the table, when the number of remaining cars is small, the probability of surviving is very small also. Indeed, the Kaplan–Meier probability reaches zero when the last car faces the wrecker.

After some summary statistics the program gives a plot of the Kaplan–Meier probability against the TIME variable. This plot is relatively inexact, even compared to most line printer plots. It is worth noting that the plotted curve does not reach the probability zero.

The following steps of analysis compare the cars of different brands with each other. First, we calculate the Kaplan–Meier estimator of the survivor function separately for each brand. Then, we generate a plot of the survivor function. Third, we statistically compare the survivor functions. Stratification is performed using the following command: STRATA MAKE=4 This command specifies that the covariate MAKE be used as stratification variable. It has four levels, the first one being indicated with 1. If the first level has index 0 the /ZERO command must be appended. The survivor tables for the four brands appear in the following printout.

```
KAPLAN-MEIER ESTIMATION
WITH STRATIFICATION ON MAKE
ALL THE DATA WILL BE USED


STRATUM NUMBER 1
```

NUMBER AT RISK	NUMBER FAILING	TIME	K-M PROBABILITY	STANDARD ERROR
8.000	1.000	69.000	0.875	0.117
7.000	1.000	76.000	0.750	0.153
6.000	1.000	80.000	0.625	0.171
5.000	2.000	87.000	0.375	0.171
3.000	2.000	90.000	0.125	0.117
1.000	1.000	101.000	0.000	

```
GROUP SIZE = 8.000
NUMBER FAILING = 8.000
PRODUCT LIMIT LIKELIHOOD = -13.863

MEAN SURVIVAL TIME = 85.000
```

FAILURE QUANTILES

75%   90.000
50%   87.000
25%   80.000

STRATUM NUMBER 2

NUMBER AT RISK	NUMBER FAILING	TIME	K-M PROBABILITY	STANDARD ERROR
8.000	1.000	66.000	0.875	0.117
7.000	1.000	76.000	0.750	0.153
6.000	1.000	80.000	0.625	0.171
5.000	1.000	88.000	0.500	0.177
4.000	1.000	89.000	0.375	0.171
3.000	1.000	90.000	0.250	0.153
2.000	1.000	101.000	0.125	0.117
1.000	1.000	109.000	0.000	

GROUP SIZE = 8.000
NUMBER FAILING = 8.000
PRODUCT LIMIT LIKELIHOOD = -16.636

MEAN SURVIVAL TIME = 87.375

FAILURE QUANTILES

75%   101.000
50%   89.000
25%   80.000

STRATUM NUMBER 3

NUMBER AT RISK	NUMBER FAILING	TIME	K-M PROBABILITY	STANDARD ERROR
8.000	1.000	34.000	0.875	0.117
7.000	1.000	54.000	0.750	0.153
6.000	1.000	60.000	0.625	0.171
5.000	1.000	65.000	0.500	0.177
4.000	1.000	73.000	0.375	0.171
3.000	1.000	88.000	0.250	0.153
2.000	2.000	96.000	0.000	

GROUP SIZE = 8.000
NUMBER FAILING = 8.000
PRODUCT LIMIT LIKELIHOOD = -15.249

MEAN SURVIVAL TIME = 70.750

FAILURE QUANTILES

75%   96.000
50%   73.000
25%   60.000

STRATUM NUMBER 4

NUMBER AT RISK	NUMBER FAILING	TIME	K-M PROBABILITY	STANDARD ERROR
8.000	1.000	69.000	0.875	0.117
7.000	1.000	89.000	0.750	0.153
6.000	2.000	99.000	0.500	0.177
4.000	1.000	102.000	0.375	0.171

```
        3.000          1.000         110.000         0.250          0.153
        2.000          1.000         121.000         0.125          0.117
        1.000          1.000         130.000         0.000

GROUP SIZE = 8.000
NUMBER FAILING = 8.000
PRODUCT LIMIT LIKELIHOOD = -15.249

MEAN SURVIVAL TIME = 102.375

FAILURE QUANTILES
-----------------
75%          121.000
50%          102.000
25%           99.000
```

The printout contains, for each brand, a survivor table parallel to the one for the entire sample. Because of the small sample sizes, the standard errors are much bigger than for the entire sample. The summary statistics suggest that the mean survival times range from 70.75 months for car from brand 3 to 102.38 months for cars from brand 4. While an inspection of the survivor functions suggests that the curves of the four brands are very similar in shape at the beginning, the brands seem to differ after the first three cars in each group stopped functioning.

The following printout displays a summary plot of the stratified survival analysis. To show another plot option provided by SURVIVAL we inserted HPLOT CHAZ before the STRATA MAKE=4 command. This command generates a cumulative hazard curve instead of the probability plot above. The resulting plot appears below, followed by the probability plot generated in another run.

```
KAPLAN-MEIER ESTIMATION

                         CUMULATIVE HAZARD
            .0000     1.5000     3.0000     4.5000     6.0000     7.5000
TIME       ·ÔöáááááááááÔáááááááááááÔáááááááááááÔáááááááááááÔáááááááááÔç
  34.000 û 3                                                         ç
         ·  3                                                        ·
         ·  3                                                        ·
         ·  3                                                        ·
         ·    3                                                      ·
  66.000 û        3                                                  ç
         ·  2   3                                                    ·
         ·  42    3                                                  ·
         ·  4 2   3                                                  ·
         ·   4     3    1                                            ·
  98.000 û   4       2    1                                          ç
         ·        4    2                                             ·
         ·          4    2                                           ·
         ·          4    2                                           ·
         ·              4                                            ·
 130.000 û             4                                             ç
           âêáááááááááááâêáááááááááááâêáááááááááááâêáááááááááááâêáááááááááááâêáááááááááááâêáááááááááááâêáááááááááááâêáááááááááááâêáááááááááááâêáááááááááááâêáì
            .0000     1.5000     3.0000     4.5000     6.0000     7.5000

PLOT TYPE:   CUMULATIVE HAZARD
HPLOT PARAMETERS:             TMIN =       34.000
        DELT =       0.000    TMAX =      130.000
        PMIN =       0.005    PMAX =        5.298
        NLINES =      15       ADJUST
```

PMIN AND PMAX ABOVE ARE TRANSFORMED.

STRATIFICATION ON MAKE SPECIFIED, 4 LEVELS

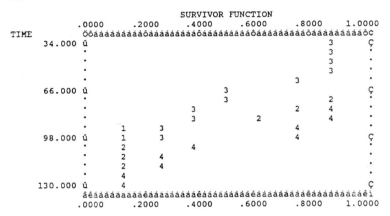

```
KAPLAN-MEIER ESTIMATION
WITH STRATIFICATION ON MAKE
ALL THE DATA WILL BE USED
                              SURVIVOR FUNCTION
                .0000    .2000    .4000    .6000    .8000   1.0000
TIME        ÖôáááááááááÔáááááááááÔáááááááááÔáááááááááÔáááááááááôÇ
      34.000 û                                            3      Ç
             .                                             3     .
             .                                             3     .
             .                                             3     .
             .                                        3          .
      66.000 û                   3                              Ç
             .                   3                      2        .
             .             3                       2    4        .
             .             3                  2         4        .
             .       1     3                       4            .
      98.000 û       1     3                       4           Ç
             .       2           4                             .
             .       2     4                                  .
             .       2     4                                  .
             .       4                                        .
     130.000 û       4                                        Ç
            âêáááááááaáâêááááááááááâêááááááááááâêáááááááááâêááááááááâáêì
                .0000    .2000    .4000    .6000    .8000   1.0000
```

The cumulative hazard function suggests that cars from brand 3 have the lowest cumulative survival probabilities, and cars from brand 4 have the highest probabilities. The survivor function shows the same pattern; however, the image is mirrored. It suggests that the highest survival probabilities can be measured for all cars at relatively early times. The brands differ, though, in that these times are shortest for cars from the third brand and longest for cars from the fourth brand. In both figures, cars from brands 1 and 2 lie between cars from brands 3 and 4, with cars from brand 2 a little longer lived than cars from brand 1.

One can test whether the differences between the strata are significant. The command LRANK/ALL yields three variations of the log-rank test. The result appears below.

```
LOG-RANK TEST, STRATIFICATION ON MAKE , STRATA RANGE 1 TO 4

                       METHOD: MANTEL
           CHI-SQ STATISTIC:        12.257 WITH   3 DOF
SIGNIFICANCE LEVEL (P VALUE):        0.007

                       METHOD: BRESLOW-GEHAN
           CHI-SQ STATISTIC:        10.856 WITH   3 DOF
SIGNIFICANCE LEVEL (P VALUE):        0.013

                       METHOD: TARONE-WARE
           CHI-SQ STATISTIC:        11.494 WITH   3 DOF
SIGNIFICANCE LEVEL (P VALUE):        0.009
```

The first test is the Mantel–Haenszel test, well-known as the log-rank test. The Breslow–Gehan statistic weights each observation time by the number of cases at risk at that time. As a result, earlier times have a greater weight than later times. The Tarone–Ware statistic weights by the square root of the number of cars at risk at a given time. This also results in a greater weight for the earlier observation periods. In the present example, all three tests allow us to reject the null hypothesis that the survival curves are the same for cars from the four brands.

The SURVIVAL program does not offer an option for planned comparisons. Thus, the three tests must be interpreted as overall tests just like main effect tests in ANOVA. An approximation of contrasts may be realized by performing pairwise tests. Here, Bonferroni adjustment of the experiment-wise alpha is strongly recommended. It helps to avoid inflation of the alpha error rate.

If the survival curves cross, the overall tests can be misleading; therefore, SURVIVAL offers as an alternative a log-rank test that focuses on subsets of the time axis. The user may specify a time point as an upper bound for the segment of the life table to be evaluated. Accordingly, a lower bound can be specified.

The SURVIVAL program is a very flexible tool for survival analysis. It speaks the same language as SYSTAT. Therefore, SURVIVAL will be easy to use for researchers familiar with SYSTAT. However, SURVIVAL does not cover the complete array of methods for event history analysis. Other estimation routines are available in BMDP (Blossfeld et al., 1985), GLIM (Baker and Nelder, 1978), and Tuma's program RATE (see Blossfeld et al., 1988, ch. 6). Sample listings for SAS, GLIM, and BMDP examples appear in Allison (1984).

Chapter 7   Growth-Curve Estimation

1. Summary of Method

Any function that attempts to describe a process that has been been measured repeatedly can be considered a growth curve (Rao, 1958; Rogosa *et al.*, 1982). Burchinal and Appelbaum (1987) describe three dimensions along which strategies for estimating growth curves can be defined.

For any process, the parameters of a growth-curve function could be calculated for each individual considered. If the parameters are expected to differ across individuals, the process has been described as a "weak developmental process" (Nesselroade and Baltes, 1979). A "strong developmental process" requires that a single set of parameters describe the process. The characteristic curve is the same for all, and individual differences appear as uniform variations around the prototypical curve (Wohlwill, 1973).

A second dimension is determined by whether the process can be classified *a priori* or must be estimated from data. If the former is the case, a family of curves provides the basis for selecting the growth-curve type. The values of the parameters can then be estimated from sample data.

The model selected for estimation represents a third dimension for parameter estimation (Burchinal and Appelbaum, 1987). Available models include polynomial growth curves (Burchinal and Appelbaum, 1987), com-

119

ponent models (Tisak and Meredith, 1990; Tucker, 1966), and various nonlinear models (Thissen and Bock, 1990).

This chapter will demonstrate two generally used methods for estimating growth curves: component analysis (Tuckerizing) and polynomial curve fitting.

Component analysis (Tucker, 1966; Tisak and Meredith, 1990) looks at the covariation across occasion of a single variable measured on a number of individuals. A higher degree of covariation between two occasions suggests that the rank order of individuals does not change but suggests nothing about changes in level. Components analysis considers the cross-products matrix, which includes information about level along with the covariation.

To estimate growth curves, component analysis computes a principal components analysis (PCA) on the cross-products matrix. The pattern of this analysis yields a set of reference curves (each component represents a single curve). The parameter estimates that define the curves are orthogonal, which, in part, means that each curve carries different information about the input matrix. It also suggests that the curves are additive.

As in ordinary PCA, one goal is often to determine the fewest number of components needed to estimate the observed data adequately. In components analysis this translates to using the minimal number of reference curves to explain the data.

Like PCA, the first solution out of the computer may not be the best. To improve interpretation of the solution, rotation of the reference curves is often suggested (Tucker, 1966; Wood, 1990). Unlike many exploratory strategies that use somewhat automatic methods for rotation, component analysis requires rotation to a set of reference curves that in some sense look like growth and make sense. In the example to be presented, we will show that an ill-advised rotation could predict a nest with a negative number of eggs. Components analysis may require "hand-rotation" that may not be readily available in a package. Matrix algebra techniques are often necessary to complete a satisfactory rotation.

The second general type of model we will demonstrate is the polynomial growth-curve model. The estimation of the model will be accomplished via ordinary regression techniques. The trick will be to generate a set of predictor variables (polynomials) that suggest separate and important characteristics of the curve.

One major difference between these polynomial models and the components models described above is the function of the variable, time, which indicates when each observation has been measured. Unlike components analysis, polynomial growth curves include the spacing between measurement occasions as part of the functional relationship described by the model. For polynomial growth curves, time becomes an independent variable. The observations of the variable of interest are considered a function of time or some transformation of time.

We will first consider the power-series model. This model uses polynomial terms representing the occasion raised to the first, second, third, etc., powers. If only the first-order polynomial were necessary, the curve would be linear. Necessary first- and second-order polynomial terms will suggest a quadratic process over time. Higher-order polynomials will represent more turning points in the curve.

The major problem with the power-series model lies in the generally high correlation among the polynomial predictor variables. To remedy this problem, we will also consider orthogonal polynomials, each of which represents a different curve type. These will be shown to be orthogonal and equivalent to the power series in terms of the adequacy of estimation.

Before delineating these two models, we offer the following caveat. The selection of "best" model depends, in part, on the process being considered. The models we present are considered general but can often be replaced by models that more directly reflect the process being considered. Another way to think of this is that each model will generate a picture of the process. For the analysis to be appropriate, the picture should reflect the main characteristics of the process.

We now present methods for computing a component model and a polynomial growth-curve model.

2. Outline of the Computational Procedures

Component analysis can make use of any program that does unrotated, unstandardized principal components analysis. We selected SAS PROC PRINCOMP (SAS Statistics, 1985). Unlike many packaged routines (e.g., SAS PROC FACTOR), PRINCOMP will not standardize across rows of the solution. With the input of a cross-products matrix, the unstandardized solution is essential to maintain the information in the original matrix.

PRINCOMP allows the user to specify the number of components to be printed, the type of matrix to be analyzed, whether or not the intercept is to be used in the equation.

Input to the program can take the form of raw data called into a SAS data base. Matrices can be read in directly, provided that the SAS file created is of the right type (to be discussed below).

3. Annotated Data Examples

Component Model

The data example is selected from a study of reproduction in the California gull (Pugesek and Wood, 1989). A sample of 176 gulls was banded and followed beginning in 1980 and through 1988. The nest of each individual was observed, and clutch size (the maximum number of eggs in the nest) was measured. Fledgling success (the number of surviving chicks) was also measured.

The clutch size, X_{it}, was assumed to be an additive sum of some set of underlying reference curves with the clutch size for any bird in a given year being

$$X_{it} = b_c Y_{it} \qquad (1)$$

where Y_{it} represents the individual's component score for curve F_i and b_c represents the component weight for component c.

Step 1: Preparing the Data

The type of data matrix to be analyzed for the component model is an $N \times T$ raw data matrix in which N is the number of cases and T is the number of trials per case. The raw data are used to compute the $T \times T$ matrix of crossproducts necessary for the analysis. This matrix can then be analyzed by any program that performs an unstandardized principal components analysis.

Since the cross-products matrix was already available, we used a SAS TYPE=COV input data step to load the cross-products matrix. The SAS input statements are

```
OPTIONS NOCENTER LINESIZE=80 NONUMBER NODATE;
DATA COVAR(TYPE=COV);
_TYPE_='COV';
INPUT _NAME_ $ N80 N84 N85 N86 N87 N88;
CARDS;
N80 311.0 173.0 302.0 168.0 253.0 278.0
N84 173.0 377.0 291.0 250.0 377.0 373.0
N85 203.0 291.0 583.0 368.0 568.0 584.0
N86 168.0 250.0 368.0 401.0 466.0 448.0
N87 253.0 377.0 568.0 466.0 867.0 761.0
N88 278.0 373.0 584.0 448.0 761.0 899.0
;
```

Step 2: Unrotated Components Analysis

SAS PROC PRINCOMP can be used to perform the unrotated unstandardized principal components analysis. The statements required are:

```
PROC PRINCOMP COV N=6 DATA=COVAR NOINT;
    VAR N80 N84 N85 N86 N87 N88;
```

Here we are requesting that a covariance matrix (actually the matrix of cross-products) be analyzed. The pattern to be produced will be unstandardized. (Note: When given a covariance matrix, PROC FACTOR will by default produce a solution in which the rows of pattern coefficients have been standardized so that the squared sums of the a set of row coefficients equals 1. This removes information regarding the variances and means from the component loadings).

The solution generated first yields a set of eigenvalues for the cross-products matrix. Tucker (1966) has suggested that these eigenvalues be tested using the Mean Square Ratio (MSR) for each factor. This test makes use of the fact that the sum of the squares of the observed scores equals the sum of the squared eigenvalues. The sum of squares of eigenvalues for the first k components equals the sum of the squares of estimates of the observed values based on using only those components. As a result, the sum of eigenvalues for components $k+1$ through t represents the sum of the squared errors of approximation when k components are used to estimate the observed scores. The MSR for component k is

$$\text{MSR}_k = \frac{\lambda^2 \ (n\text{-}k)(N\text{-}k)}{\sum_{k+1}^{t} \lambda^2 \ (n+N+1\text{-}2k)} \tag{2}$$

This statistic tends to be upwardly biased when compared with the F distribution (Tucker, 1966). As a result, it is best used as a descriptive indication of the importance of a component.

Pugesek and Wood (1989) reported the MSRs for the gull data as 16.65, 1.68, 1.37, 1.28, 1.29, and 0.00. Based on the values, they selected two components. The output from the PRINCOMP procedure follows.

```
SAS
```

```
PRINCIPAL COMPONENT ANALYSIS
```

```
    10000 OBSERVATIONS
        6 VARIABLES
```

```
TOTAL VARIANCE=3438
```

	EIGENVALUE	DIFFERENCE	PROPORTION	CUMULATIVE
PRIN1	2664.44	2430.60	0.774998	0.77500
PRIN2	233.84	61.97	0.068017	0.84302
PRIN3	171.87	26.73	0.049991	0.89301
PRIN4	145.14	18.89	0.042215	0.93522
PRIN5	126.25	29.79	0.036722	0.97194
PRIN6	96.46		0.028056	1.00000

EIGENVECTORS

	PRIN1	PRIN2	PRIN3	PRIN4	PRIN5	PRIN6
N80	0.203525	0.801918	-.474152	0.148052	0.245826	-.091295
N84	0.283439	0.485563	0.659890	-.304358	-.388483	-.069879
N85	0.419366	-.152173	0.079325	0.768706	-.289412	-.346431
N86	0.334026	-.001548	0.328203	0.273498	0.458337	0.704154
N87	0.539154	-.256903	0.047369	-.367274	0.526539	-.478474
N88	0.547824	-.178880	-.472727	-.291285	-.466450	0.376825

As can be seen, PRINCOMP generated eigenvalues and eigenvectors for all six components. To suppress the printing of the last four components, we could have requested that only the first two eigenvectors be printed.

The eigenvalues and eigenvectors can be used to generate the reference-curve coefficients. For each eigenvector, a, a corresponding vector can be generated using the formula

$$p = a\,\lambda\,/\,\sqrt{N} \tag{3}$$

where λ is the square root of the eigenvalue corresponding to a and N is the sample size; p thus represents the corresponding pattern vector.

While the reference-curve coefficients could easily be calculated by hand, we include a SAS PROC IML program that will generate the coefficients. In addition, this program will provide component scores if the raw data is included as input.

```
OPTIONS NOCENTER LINESIZE=80 NONUMBER NODATE;
PROC IML;
  X={311.0 173.0 203.0 168.0 253.0 278.0,
     173.0 377.0 291.0 250.0 377.0 373.0,
     203.0 291.0 583.0 368.0 568.0 584.0,
     168.0 250.0 368.0 401.0 466.0 448.0,
     253.0 377.0 568.0 466.0 867.0 761.0,
     278.0 373.0 584.0 448.0 761.0 899.0};
EIGVALS=EIGVAL(X);
  PRINT 'EIGENVALUES',EIGVALS;
EIGVECS=EIGVEC(X);
  PRINT 'EIGENVECTORS' EIGVECS;
NOCCS={6}; NFACTS={6}; NINDS={176};
INDFACTS=1:NFACTS;
  PRINT INDFACTS;
SEIGVALS=DIAG(SQRT(EIGVALS(|INDFACTS, |)));
  PRINT 'CHARACTERISTIC ROOTS',,SEIGVALS;
SEIGVECS=EIGVECS(| ,INDFACTS|);
  PRINT 'LEFT PRINCIPAL VECTORS',SEIGVECS;
PATTERN=(NINDS##-.5)#SEIGVECS*SEIGVALS;
  PRINT 'UNROTATED FACTOR PATTERN',PATTERN;
*SCORES=(NINDS/(SEIGVALS##2))*PATTERN`*RAWDATA;
```

The output from this program follows.

EIGENVALUES

EIGVALS	COL1
ROW1	2664.4
ROW2	233.8
ROW3	171.9
ROW4	145.1
ROW5	126.2
ROW6	96.4570

EIGENVECTORS

EIGVECS	COL1	COL2	COL3	COL4	COL5	COL6
ROW1	0.2035	0.8019	-0.4742	0.1481	0.2458	-0.0913
ROW2	0.2834	0.4856	0.6599	-0.3044	-0.3885	-0.0699
ROW3	0.4194	-0.1522	0.0793	0.7687	-0.2894	-0.3464
ROW4	0.3340	-.001548	0.3282	0.2735	0.4583	0.7042
ROW5	0.5392	-0.2569	0.0474	-0.3673	0.5265	-0.4785
ROW6	0.5478	-0.1789	-0.4727	-0.2913	-0.4665	0.3768

CHARACTERISTIC ROOTS

SEIGVALS	COL1	COL2	COL3	COL4	COL5	COL6
ROW1	51.6182	0	0	0	0	0
ROW2	0	15.2919	0	0	0	0
ROW3	0	0	13.1099	0	0	0
ROW4	0	0	0	12.0473	0	0
ROW5	0	0	0	0	11.2361	0
ROW6	0	0	0	0	0	9.8213

LEFT PRINCIPAL VECTORS

SEIGVECS	COL1	COL2	COL3	COL4	COL5	COL6
ROW1	0.2035	0.8019	-0.4742	0.1481	0.2458	-0.0913
ROW2	0.2834	0.4856	0.6599	-0.3044	-0.3885	-0.0699
ROW3	0.4194	-0.1522	0.0793	0.7687	-0.2894	-0.3464
ROW4	0.3340	-.001548	0.3282	0.2735	0.4583	0.7042
ROW5	0.5392	-0.2569	0.0474	-0.3673	0.5265	-0.4785
ROW6	0.5478	-0.1789	-0.4727	-0.2913	-0.4665	0.3768

UNROTATED FACTOR PATTERN

PATTERN	COL1	COL2	COL3	COL4	COL5	COL6
ROW1	0.7919	0.9243	-0.4686	0.1344	0.2082	-0.0676
ROW2	1.1028	0.5597	0.6521	-0.2764	-0.3290	-0.0517
ROW3	1.6317	-0.1754	0.0784	0.6981	-0.2451	-0.2565
ROW4	1.2997	-.001784	0.3243	0.2484	0.3882	0.5213
ROW5	2.0978	-0.2961	0.0468	-0.3335	0.4460	-0.3542
ROW6	2.1315	-0.2062	-0.4671	-0.2645	-0.3951	0.2790

The matrix, PATTERN, is the matrix of reference-curve coefficients.

At this point we could predict the clutch size for each bird in a particular year using equation 1. Each reference curve represents a component of clutch size that can be estimated. If the reference curve is prototypical, the plot of that curve suggests how the portion of the mean clutch size estimated by the component changes from year to year. As the reference curves are orthogonal, the two reference curves suggested by the first two components may represent different processes that contribute to the estimation of clutch size. The reference curves are plotted in Figure 1 using trial as the abscissa and the component loading as the ordinate.

Step 3: A Rotated Components Solution

Once an optimal number of reference curves has been established, those curves can be rotated to come up with a substantively more interesting solution. Elegant methods developed for computerized "hand-rotation" include a SAS IML procedure (Wood, 1989) and an interactive FORTRAN proce-

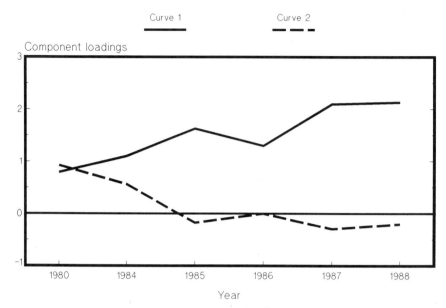

FIGURE 7.1.   Unrotated reference curves.

dure (Surra, 1988). Both can be used in the IBM CMS environment. We will present a simple method for rotating factors using SAS IML.

Most of the standard rotation procedures available in packages are inappropriate for rotating reference curves either because they are substantively indefensible (e.g., VARIMAX, ORTHOMAX, PROMAX) or because they rescale the solution losing the cross-products information (e.g., PROCRUS-TES). That means that the investigator must establish the criterion for an appropriate rotation and select a translation matrix that will do the job. Pugesek and Wood (1989) were interested in having non-negative values for the reference curves. This could be translated into the notion that a good reference curve should not indicate that a negative number of eggs should be expected in any nest. The translation matrix they selected to try was

$$T = \begin{bmatrix} .0698 & .9976 \\ .9976 & -.0698 \end{bmatrix} \qquad (4)$$

where $T = [\cos \phi \sin \phi \, , \, \sin \phi -\cos \phi \,]$ in which ϕ is the angle of rotation.
Using this matrix, the solution can be rotated by

$$b_{rot} = b * T. \qquad (5)$$

The PROC IML solution for this, along with the plot (Figure 2) of the new reference curves, follows:

```
PROC IML;
START;
F={.7919 1.1028 1.6317 1.2997 2.0978 2.1315,
   .9243 .5597 -.1754 -.0018 -.2961 -.2062};
T={.0698 .9976,.9976 -.0698};
F=F`;
PRINT F;
TRANS=F*T;
PRINT TRANS;
FINISH;
RUN;
```

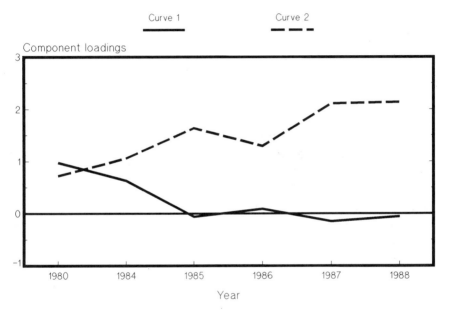

FIGURE 7.2.   Rotated reference curves.

F	COL1	COL2
ROW1	0.7919	0.9243
ROW2	1.1028	0.5597
ROW3	1.6317	-0.1754
ROW4	1.2997	-0.0018
ROW5	2.0978	-0.2961
ROW6	2.1315	-0.2062

TRANS	COL1	COL2
ROW1	0.9774	0.7255
ROW2	0.6353	1.0611
ROW3	-0.0611	1.6400
ROW4	0.0889	1.2967
ROW5	-0.1490	2.1134
ROW6	-0.0569	2.1408

To rotate the curves in this fashion requires both a good guess at the translation matrix and a substantive criterion for determining the adequacy of the rotation. As this method is basically hit or miss, users may prefer an iterative method that allows rotation to a specified criterion (Wood, 1990).

The model generates a set of clutch-size estimation scores. The correlations between actual and predicted clutch size for each year can function as a test of the adequacy of the model. Low correlations could suggest that additional components should be added to the model.

POLYNOMIAL GROWTH CURVES

Step 1: Preparing the Data

The data for this example were simulated to represent a longitudinal process that could be modeled by the equation

$$y_t = 2 + .5*t + .25*t^2 + .12*t^3. \tag{6}$$

Separate stochastic terms were added to represent variation due to individuals and variation due to occasion. The sample was divided into two groups. An overall mean-level group difference was created by adding a constant for one group. An additional stochastic term was added for the second group. The program used to create the data follows:

```
OPTIONS NOCENTER LINESIZE=80 NONUMBER NODATE;
CMS FILEDEF OUTFILE DISK ORPOLSIM DATA E1;
DATA ORPOLSIM;
SEED1=1232763476;
SEED2=423248;
SEED3=213426;
DO I=1 TO 50;
  RESIDI=3*RANUNI(SEED2);
DO T=1 TO 8;
  RESIDT=6*RANUNI(SEED1);
    Y=2+.5*T+.25*T**2+.12*T**3+RESIDT+RESIDI;
      ICHECK=I/2;  IINT=INT(I/2);
      IF ICHECK EQ IINT THEN Y=Y+RANUNI(SEED3);
      IF ICHECK NE IINT THEN Y=Y+3;
            FILE OUTFILE;
            PUT I T Y;
  END;
    END;
```

The program generated 50 cases of data with each case measured at eight occasions. A plot of the data appears in Figure 3.

The stochastic terms were generated using the RANUNI function which selects a random number from a uniform distribution. This represents a violation of the assumption of normally distributed residuals and should make the model a little more difficult to fit.

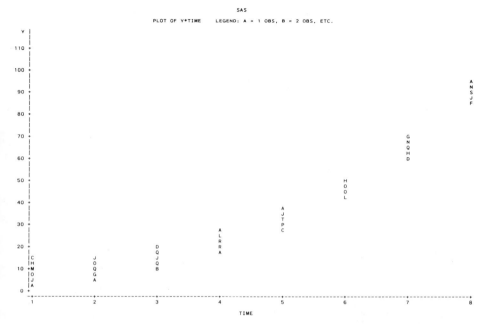

FIGURE 7.3.   Reference curves of clutch-size estimation. A = 1 OBS, B = 2 OBS, etc.

Step 2: Fitting a Polynomial Using Time-dependent (Power Series)
Coefficients

The strategy for fitting a polynomial growth curve involves selected an appropriate basis to be used in creating a design matrix which can then be used in any regression-type algorithm. We will demonstrate the use of two different kinds of bases: time-dependent polynomials and orthogonal polynomials.

A basis matrix for the time-dependent polynomials has the form

$$
\begin{array}{cccccccc}
1 & t_1 & t_1^2 & t_1^3 & t_1^4 & . & . & t_1^K \\
1 & t_2 & t_2^2 & t_2^3 & t_2^4 & . & . & t_2^K \\
1 & t_3 & t_3^2 & t_3^3 & t_3^4 & . & . & t_3^K \\
. & . & . & . & . & . & . & . \\
. & . & . & . & . & . & . & . \\
. & . & . & . & . & . & . & . \\
1 & t_t & t_t^2 & t_t^3 & t_t^4 & . & . & t_t K
\end{array}
\tag{7}
$$

where t_t can be the occasions of measurement (e.g., $t_1 = 1, t_2 = 2$, etc.) and K represents the highest-order polynomial used in predicting. The values of t should reflect the spacing between measurement occasions. These data have been simulated assuming equally spaced occasions.

The power-series form for the equation of prediction based on the above matrix could be written as

$$y_t = a + b_1 t_t + b_2 t_t^2 + b_3 t_t^3 + \ldots + b_K t_t^K. \tag{8}$$

This equation is one of a family of polynomial estimation equations. A more general equation for polynomial estimation is

$$y_t = a + b_1 p_{1t} + b_2 p_{2t} + b_3 p_{3t} + \ldots + b_K p_{Kt} \tag{9}$$

in which p_{kt} represents *the TIME value raised to the k* value.

A SAS PROC REG procedure was employed to estimate the polynomial coefficients. Since the basis vectors are not independent, their intercorrelations and correlations with the outcome variable were computed along with the regression. The program statements follow:

```
OPTIONS NOCENTER LINESIZE=80 NONUMBER NODATE;
DATA ORPOLSM2;
  INFILE CARDS;
  INPUT ID TIME Y;
    TIMESQ=TIME*TIME;
    TIMECUB=TIME*TIME*TIME;
  ICHECK=ID/2;
  IINT=INT(ID/2);
  GROUP=1;
  IF IINT EQ ICHECK THEN GROUP=2;
CARDS;
<CARDS>
;
PROC REG;
  MODEL Y=TIME TIMESQ TIMECUB;
PROC CORR;
  VAR Y TIME TIMESQ TIMECUB;
PROC PLOT;
  PLOT Y*TIME;
PROC SORT;
  BY GROUP;
PROC REG;
  BY GROUP;
  MODEL Y=TIME TIMESQ TIMECUB;
```

This procedure produced the following output:

```
SAS

DEP VARIABLE: Y
ANALYSIS OF VARIANCE
```

| | | SUM OF | MEAN | | |
SOURCE	DF	SQUARES	SQUARE	F VALUE	PROB>F
MODEL	3	291836.91	97278.97133	15790.093	0.0001
ERROR	396	2439.66094	6.16075996		
C TOTAL	399	294276.57			

ROOT MSE	2.482088	R-SQUARE	0.9917
DEP MEAN	36.36695	ADJ R-SQ	0.9916
C.V.	6.825119		

PARAMETER ESTIMATES

VARIABLE	DF	PARAMETER ESTIMATE	STANDARD ERROR	T FOR HO: PARAMETER=0	PROB > ¦T¦
INTERCEP	1	7.84513897	0.86492328	9.070	0.0001
TIME	1	0.70248891	0.78262710	0.898	0.3699
TIMESQ	1	0.23570067	0.19631120	1.201	0.2306
TIMECUB	1	0.11944596	0.01440253	8.293	0.0001

As can be seen the regression reproduced the coefficients for the quadratic and cubic polynomial terms. The estimates of the intercept and linear term were somewhat larger than the simulation term. This was anticipated, as the general increase over time was exaggerated due to the overall level difference for the two groups.

The simulation created two groups with the same developmental function. The two groups differed in overall mean level and in variation. To consider the differences we ran the regressions separately for both groups. The results of the regression were

SAS

GROUP=1

DEP VARIABLE: Y
ANALYSIS OF VARIANCE

SOURCE	DF	SUM OF SQUARES	MEAN SQUARE	F VALUE	PROB>F
MODEL	3	146227.31	48742.43651	12735.900	0.0001
ERROR	196	750.12505	3.82716860		
C TOTAL	199	146977.43			

ROOT MSE	1.956315	R-SQUARE	0.9949	
DEP MEAN	37.93075	ADJ R-SQ	0.9948	
C.V.	5.157597			

PARAMETER ESTIMATES

VARIABLE	DF	PARAMETER ESTIMATE	STANDARD ERROR	T FOR HO: PARAMETER=0	PROB > ¦T¦
INTERCEP	1	9.34462349	0.96408259	9.693	0.0001
TIME	1	0.87710285	0.87235155	1.005	0.3159
TIMESQ	1	0.16728101	0.21881734	0.764	0.4455
TIMECUB	1	0.12576230	0.01605372	7.834	0.0001

SAS

GROUP=2

DEP VARIABLE: Y
ANALYSIS OF VARIANCE

SOURCE	DF	SUM OF SQUARES	MEAN SQUARE	F VALUE	PROB>F
MODEL	3	145613.18	48537.72555	13441.113	0.0001
ERROR	196	707.78322	3.61113886		
C TOTAL	199	146320.96			

```
        ROOT MSE        1.9003    R-SQUARE      0.9952
        DEP MEAN       34.80316   ADJ R-SQ      0.9951
        C.V.            5.460136
```

PARAMETER ESTIMATES

VARIABLE	DF	PARAMETER ESTIMATE	STANDARD ERROR	T FOR H0: PARAMETER=0	PROB > |T|
INTERCEP	1	6.34565445	0.93647791	6.776	0.0001
TIME	1	0.52787496	0.84737341	0.623	0.5340
TIMESQ	1	0.30412033	0.21255192	1.431	0.1541
TIMECUB	1	0.11312961	0.01559405	7.255	0.0001

Although the higher-order coefficients of this model are reproduced, the terms of the regression are highly correlated. The degree of correlation is

VARIABLE	N	MEAN	STD DEV
Y	400	36.36695130	27.15760070
TIME	400	4.50000000	2.29415734
TIMESQ	400	25.50000000	21.15108558
TIMECUB	400	162.00000000	172.33906081

PEARSON CORRELATION COEFFICIENTS / PROB > |R| UNDER H0:RHO=0 / N = 400

	Y	TIME	TIMESQ	TIMECUB
Y	1.00000	0.94486	0.99010	0.99458
	0.0000	0.0001	0.0001	0.0001
TIME	0.94486	1.00000	0.97619	0.93183
	0.0001	0.0000	0.0001	0.0001
TIMESQ	0.99010	0.97619	1.00000	0.98761
	0.0001	0.0001	0.0000	0.0001
TIMECUB	0.99458	0.93183	0.98761	1.00000
	0.0001	0.0001	0.0001	0.0000

With this high degree of degree of collinearity, the standard errors of the tests of the estimates tend to become unstable (Neter, *et al.*, 1985). An alternative strategy might then be to select a set of basis vectors that are independent.

If parsimony is a consideration, the predictive model with the fewest terms may be of interest. A number of hierarchical procedures can be imagined. One possibility is to begin with linear polynomial and test the fit. Higher-order polynomials can be added to the equation with the increment in fit an indication of whether the added term improves the model. If this strategy is used, one cannot assume that when a power is found that does not contribute to the explanation of variation, all higher-order power terms will contribute nothing to the explanation of variation. The higher order powers would still have to be tested to make this determination.

On the other extreme, if a particular process (e.g., a second-order power series) is hypothesized, that process can be tested directly.

As higher-order powers are added to improve interpolation, extrapolation becomes more dangerous with predicted values quickly heading off in the general direction of infinity.

Step 3: Fitting a Model Using Orthogonal Polynomials

The prediction equation for orthogonal polynomials is

$$y_t = a + b_1\, p_{1t} + b_2\, p_{2t} + b_3\, p_{3t} + b_4\, p_{4t} + \ldots + b_k\, p_{kt}, \qquad (10)$$

where p_{kt} is the polynomial coefficient corresponding to a polynomial of order, k, for occasion, t, and K is highest-order polynomial used in estimating the curve.

The basis matrix for eight occasions of measurement using orthogonal polynomials as estimators is

a	lin	quad	cub	quart	quint
1	−7	7	−7	7	−7
1	−5	1	5	−13	23
1	−3	−3	7	−3	−17
1	−1	−5	3	9	−15
1	1	−5	−3	9	15
1	3	−3	−7	−3	17
1	5	1	−5	−13	−23
1	7	7	7	7	7.

(11)

(See Kirk, 1982, Table E-12). This matrix would assume that the fifth-order term is the highest polynomial term to be included in the curve estimation. The first row in this table represents the values of p for occasion 1 if up to a fifth-order polynomial is used in the estimation; the second row, occasion 2, and so on through the last row representing occasion 8. These values function as variable values in the prediction equation with the weights, b_k, representing regression weights. Whereas the previous analysis defined the polynomial terms as powers of TIME, this analysis uses a different set of the polynomials—values reflecting the degree to which a particular trend type predicts a value on a specific occasion.

The following SAS data step was used to set up the input the data and create the polynomials:

```
OPTIONS NOCENTER LINESIZE=80 NONUMBER NODATE;
DATA ORPOLSM2;
   INFILE CARDS;
   INPUT ID TIME Y;
IF TIME EQ 1 THEN DO;
   LIN=-7; QUAD=7; CUB=-7; QUART=7; QUINT=-7;
   END;
IF TIME EQ 2 THEN DO;
   LIN=-5; QUAD=1; CUB=5; QUART=-13; QUINT=23;
   END;
```

```
IF TIME EQ 3 THEN DO;
   LIN=-3; QUAD=-3; CUB=7; QUART=-3; QUINT=-17;
   END;
IF TIME EQ 4 THEN DO;
   LIN=-1; QUAD=-5; CUB=3; QUART=9; QUINT=-15;
   END;
IF TIME EQ 5 THEN DO;
   LIN=1; QUAD=-5; CUB=-3; QUART=9; QUINT=15;
   END;
IF TIME EQ 6 THEN DO;
   LIN=3; QUAD=-3; CUB=-7; QUART=-3; QUINT=17;
   END;
IF TIME EQ 7 THEN DO;
   LIN=5; QUAD=1; CUB=-5; QUART=-13; QUINT=-23;
   END;
IF TIME EQ 8 THEN DO;
   LIN=7; QUAD=7; CUB=7; QUART=7; QUINT=7;
   END;
  ICHECK=ID/2;
  IINT=INT(ID/2);
  GROUP=1;
  IF IINT EQ ICHECK THEN GROUP=2;
CARDS;
<CARDS>
;
PROC REG;
  MODEL Y=LIN;
  MODEL Y=LIN QUAD;
  MODEL Y=LIN QUAD CUB;
  MODEL Y=LIN QUAD CUB QUART;
  MODEL Y=LIN QUAD CUB QUART QUINT;
PROC CORR;
  VAR Y LIN QUAD CUB QUART QUINT;
PROC PLOT;
  PLOT Y*TIME;
PROC SORT;
  BY GROUP;
PROC REG;
  BY GROUP;
  MODEL Y=LIN;
  MODEL Y=LIN QUAD;
  MODEL Y=LIN QUAD CUB;
  MODEL Y=LIN QUAD CUB QUART;
  MODEL Y=LIN QUAD CUB QUART QUINT;
```

The new set of independent variables is LIN, QUAD, CUB, etc. The results of the regression using three polynomials as predictors follows:

```
SAS

DEP VARIABLE: Y
ANALYSIS OF VARIANCE
```

		SUM OF	MEAN		
SOURCE	DF	SQUARES	SQUARE	F VALUE	PROB>F
MODEL	3	291836.91	97278.97133	15790.093	0.0001
ERROR	396	2439.66094	6.16075996		
C TOTAL	399	294276.57			

ROOT MSE	2.482088	R-SQUARE	0.9917
DEP MEAN	36.36695	ADJ R-SQ	0.9916
C.V.	6.825119		

```
PARAMETER ESTIMATES
```

VARIABLE	DF	PARAMETER ESTIMATE	STANDARD ERROR	T FOR HO: PARAMETER=0	PROB > |T|
INTERCEP	1	36.36695130	0.12410439	293.035	0.0001
LIN	1	5.59250599	0.02708180	206.504	0.0001
QUAD	1	1.84822110	0.02708180	68.246	0.0001
CUB	1	0.17916894	0.02160380	8.293	0.0001

The results of the analysis are essentially the same as the prior analysis in terms of the variance explained. The advantage of this strategy is indicated in the pattern of correlations among these predictors:

```
VARIABLE        N       MEAN      STD DEV

Y             400   36.36695     27.15760
LIN           400    0.00000      4.58831
QUAD          400    0.00000      4.58831
CUB           400    0.00000      5.75176
QUART         400    0.00000      8.78595
QUINT         400    0.00000     16.54340


PEARSON CORRELATION COEFFICIENTS / PROB > |R| UNDER HO:RHO=0 / N = 400

                Y        LIN       QUAD       CUB       QUART      QUINT


Y          1.00000   0.94486   0.31226    0.03795  -0.00479  -0.00500
           0.0000    0.0001    0.0001     0.4492    0.9239    0.9207

LIN        0.94486   1.00000   0.00000    0.00000   0.00000   0.00000
           0.0001    0.0000    1.0000     1.0000    1.0000    1.0000

QUAD       0.31226   0.00000   1.00000    0.00000   0.00000   0.00000
           0.0001    1.0000    0.0000     1.0000    1.0000    1.0000

CUB        0.03795   0.00000   0.00000    1.00000   0.00000   0.00000
           0.4492    1.0000    1.0000     0.0000    1.0000    1.0000

QUART     -0.00479   0.00000   0.00000    0.00000   1.00000   0.00000
           0.9239    1.0000    1.0000     1.0000    0.0000    1.0000

QUINT     -0.00500   0.00000   0.00000    0.00000   0.00000   1.00000
           0.9207    1.0000    1.0000     1.0000    1.0000    0.0000
```

These predictors are orthogonal. The regression weights represent the degree to which the growth curve is linear, quadratic, and cubic, respectively. This analysis separated orthogonal characteristics of the growth curve suggesting the importance of linear, quadratic, and cubic trends in estimating the data.

```
GROUP=1

DEP VARIABLE: Y
ANALYSIS OF VARIANCE

                    SUM OF       MEAN
SOURCE      DF      SQUARES      SQUARE      F VALUE      PROB>F

MODEL        3    146227.31   48742.43651   12735.900     0.0001
ERROR      196       750.12505  3.82716860
C TOTAL    199    146977.43

        ROOT MSE     1.956315    R-SQUARE     0.9949
        DEP MEAN    37.93075     ADJ R-SQ     0.9948
        C.V.         5.157597

PARAMETER ESTIMATES

                    PARAMETER     STANDARD     T FOR HO:
VARIABLE    DF      ESTIMATE      ERROR        PARAMETER=0    PROB > |T|

INTERCEP     1     37.93074517   0.13833236    274.200         0.0001
LIN          1      5.59299658   0.03018660    185.281         0.0001
QUAD         1      1.86507210   0.03018660     61.785         0.0001
CUB          1      0.18864345   0.02408057      7.834         0.0001
```

```
GROUP=2

DEP VARIABLE: Y
ANALYSIS OF VARIANCE

                   SUM OF        MEAN
SOURCE      DF     SQUARES       SQUARE      F VALUE      PROB>F

MODEL        3    145613.18    48537.72555   13441.113    0.0001
ERROR      196     707.78322    3.61113886
C TOTAL    199    146320.96

      ROOT MSE       1.9003     R-SQUARE     0.9952
      DEP MEAN      34.80316    ADJ R-SQ     0.9951
      C.V.          5.460136

PARAMETER ESTIMATES

                   PARAMETER    STANDARD     T FOR HO:
VARIABLE    DF     ESTIMATE      ERROR      PARAMETER=0   PROB > |T|

INTERCEP     1    34.80315743  0.13437148   259.007       0.0001
LIN          1     5.59201541  0.02932226   190.709       0.0001
QUAD         1     1.83137010  0.02932226    62.457       0.0001
CUB          1     0.16969442  0.02339107     7.255       0.0001
```

Separate subgroup analyses are also reported. Notice that the pattern of parameter estimates is essentially the same for both groups. This was not the case for the power-series estimates.

4. Discussion

Tucker analysis and polynomial curve fitting represent only two of the many ways in which a growth curve can be estimated. Both tend to be used in an exploratory fashion for estimating growth curves. If a stronger hypothesis exists both can be modified to test a hypothesis. For example, difference curves for different groups can be easily characterized by estimating separate subgroup models and testing for differences in the parameters. In the presence of stronger hypotheses, somewhat more confirmatory methods have been suggested (McArdle and Epstein, 1987). Other models have been suggested for estimating growth curves. In particular non-linear models (Thissen and Bock, 1990) have been suggested for modeling physical growth. Learning models are often estimated using an S-shaped response function such as a logit.

   If the hypothesis suggested includes the shape of the growth curve, that particular function probably represents the most direct test. If the model is exploratory but assumes a single curve type, higher-order polynomials will generally do a good job. If the curve modeled appears cyclical, the set of trigonometric polynomials (Fourier series) may represent a good choice.

Chapter 8   Spectral Analysis

1. Summary of Method

Spectral analysis is a method for time-series analysis in the *frequency domain*. Methods in this domain decompose a time series into a sum of processes with parameters a_i. These parameters are uncorrelated, have an expectancy of zero and an expected variance of σ^2. Examples of decompositions include representations as waves that can be described using trigonometric functions, for instance, sine functions (Brockwell and Davis, 1987; Larsen, 1990).

Alternatives to methods for analysis of time series in the frequency domain are methods for the analysis in the *time domain*. These methods use autocovariance functions to depict time series as a function of earlier points in the same series or as a trend in time.

Spectral analysis is a method useful in particular when the researcher assumes a time series of observations is determined by a number of underlying, cyclical functions. This number is typically relatively small, and each of the functions has a substantive interpretation. Examples include circadian rhythms of physiological variables such as blood pressure and psychological variables such as fatigue, attention span, and mood.

137

Spectral analysis and other methods for analysis in the frequency domain are particularly useful for three reasons:

1. they allow one to decompose a time series in underlying processes that differ in major characteristics as length of cycle,

2. they are custom-tailored for the analysis of cyclical processes, that is, processes that other methods—e.g., trend analysis with polynomial decomposition as in analysis of variance—can depict only with difficulties, and

3. they allow the researcher to analyze long time series, that is series with 32 or more occasions. (Notice that, in theory, shorter time series can also be decomposed.)

Results of spectral analysis are typically expressed in terms of a *periodogram*, or Fourier transform. The concept of a periodogram can be explained using the following analogy. In linear regression, we approximate a function using a straight time, or a first-order polynomial. The well-known equation for linear regression is

$$y_t = a_0 + a_1 x_t + e_t \tag{1}$$

where t indexes the observation points. In curvilinear regression, we approximate a function using a curved line, or a second- or higher-order polynomial. An example of curvilinear regression is the following equation for cubic regression:

$$y_t = a_0 + a_1 x_t + a_2 x_t^2 + a_3 x_t^3 + e_t = a_0 + \sum_i a_i x_t^i + e_t. \tag{2}$$

A Fourier series is defined by an approximation using weighted sine and cosine functions. The model equation is

$$y_t = a_0 + \sum_{i+1}^{q} (a_i c_{it} + b_i s_{it}) \tag{3}$$

where $c_{it} = \cos(2\pi f_i t)$, $s_{it} = \sin(2\pi f_i t)$, f_i is the ith cycle in T, or $f_i = i/T$, and $i = 1, ..., T$. The number q is determined by T. We obtain

$$q = \begin{cases} T/2 & \text{if T is even} \\ (T\text{-}2)/2 & \text{if T is odd.} \end{cases} \tag{4}$$

As in regression analysis, the parameters a_i and b_i can be estimated using the ordinary least square approach. The *periodogram* is given by the equation

$$I(f_i) = \frac{T}{2} (a_i^2 + b_i^2), \text{ for } i = 1, ..., q. \tag{5}$$

If, $(a_i^2 + b_i^2)^{1/2} = 0$, or the estimated $(\hat{a}_i^2 + \hat{b}_i^2)^{1/2}$ approximates zero, then the frequency $f_i = i/T$ is not present in the observed time series. The periodogram expresses the intensity of a wave in terms of the frequency f_i (*cf.* Brillinger,

1975; Brockwell and Davis, 1987; Davies, 1983; Krause and Metzler, 1984).
Another characteristic of the periodogram is that

$$\sum_{i=1}^{q} I(f_i) = \sum_{t=1}^{T} (y_i - \bar{y})^2, \qquad (6)$$

or, in words, the periodogram constitutes a variance decomposition of a
time series. $I(f_i)$ gives the sum of the squared deviations from the mean that
is due to the wave with frequency f_i. When analyzing periodograms, we are
interested in those frequencies that cause the largest values of $I(f_i)$, or local
maxima. Local maxima can be caused by cyclical processes but also by random factors; therefore, in many instances significance tests comparing the
relative magnitude of the $I(f_i)$ can be useful. The following test may be used
(Metzler and Nickel, 1986). The null hypothesis is as follows: The time series is a sequence of independent, normally distributed random variables
with equal means and variances. To test whether a given $I(f_i)$ is greater than
tenable under this null hypothesis, we calculate

$$F = (n-1) I(f_i)/(\Sigma_j I(f_j)) \qquad \text{for } j \neq i \qquad (7)$$

where n denotes the sample size, or the number of observation points.
Under the null hypothesis F follows an F-distribution with 2 and n-2 degrees
of freedom.

2. Overview of the Computational Procedures

This section overviews the steps for computing a periodogram using a PC.
In the next section we will illustrate these steps using an artificial data example. The SERIES module in the SYSTAT package will be used.

The SERIES module in SYSTAT requires data from a system file. For
many applications of spectral analysis, data need not be transformed before
analysis. Often, however, it seems useful to remove the mean from a time series, to replace the values with their logarithms, or to square the values. One
removes the mean to make time series comparable in their shape, disregard-
ing level differences. One substitutes the logarithms for the original values
to remove nonstationary variability. Squaring values is useful to get more
pronounced peaks in the time series. One transformation that is strongly
recommended is to remove the trend of a time series. Spectral analysis assumes trendless time series.

The SYSTAT SERIES module reads data from a SYSTAT system file. The
program expects just the time series as input information. The SERIES
module contains several options for time-series analysis. Spectral analysis is
started by typing FOURIER and the variable name.

3.   An Annotated Data Example

The following example analyzes artificial data. A stationary series of measures was generated using the following function:

$$y = 2 \sin(t) + \cos (0.5t) + \sin (\cos(65t)) + RAND \tag{8}$$

where t denotes the number of observation points; for the present example
we chose $t = 64$. $2 \sin(t)$ is a function that has both 10 maxima and 10 minima between 1 and 64. Cos $(0.5t)$ is a function with both 5 maxima and 5
minima in the interval between 1 and 64. The third trigonometric function,
sin $(\cos(65t))$, has two maxima and two minima in the interval between 1
and 64. However, between these maxima and minima, it displays 22 local
maxima and 22 local minima. Thus it is a curve oscillating in sharp ups and
downs.

The curve was generated using LOTUS 1–2–3. This program interprets
the arguments in its trigonometric functions as degrees. Therefore, we obtain, for instance, sin $(1) = 0.84$. (Notice that interpreting 1 as a real valued
number, we obtain sin $(1) = 0.02$.).

RAND is the random number generator provided by LOTUS 1–2–3, release 1A. The generator yields a trendless random number between 0 and 1.
This number was added to transform the deterministic time-series variable
into a random variable.

The time series under analysis resulted from adding the three trigonometric functions and the random numbers. Figure 1 contains this curve.

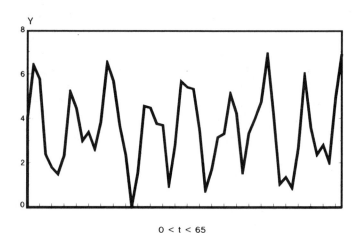

0 < t < 65

FIGURE 8.1.   Simulated   time   series. $Y = 2*\sin(t) + \cos(0.5*t) + \sin(\cos(65*t)) + RAND + 5$.

Step 1:   Data Preparation

The data were generated using LOTUS 1-2-3, release 1A. It would have been possible to generate just the curve shown in Figure 1. However, to be able to describe the components that contribute to this curve we used one column for *t*, one column each for the trigonometric functions, one column for the random numbers, and one column for the sum of the components. Thus a total of six variables resulted. The following commands are needed to send this worksheet to a file:

/Print (Initiates the output.)

File (Programs prompts for file name.)

Range (Initiates the specification of the cell range to be sent to a file.)

A1.F64 (Specifies the cell range for the present example.)

Go (Starts data transmission. Only the numbers in the cells are saved, but
    not the cell formulas.)

The file generated by LOTUS 1-2-3 contains ASCII code data that can directly be read from SYSTAT, version 4.0. For earlier versions of SYSTAT some preparatory steps are needed (See, for instance, p. 27 in the manual for SYSTAT, version 2.0). Notice that the file generated by LOTUS 1-2-3 has the suffix .PRN. Suppose the file for our example has the name FOUR.PRN. The following DOS command renames it such that SYSTAT can read it as an ASCII raw data file: RENAME FOUR.PRN FOUR.DAT. The resulting raw data are given below.

```
 1  1.682941  0.877582  -0.53326  0.147506  2.174767
 2  1.818594  0.540302  -0.35908  0.414100  2.413908
 3  0.282240  0.070737   0.828051 0.690568  1.871596
 4 -1.51360  -0.41614   -0.66701  0.895124 -1.70164
 5 -1.91784  -0.80114   -0.15361  0.006453 -2.86614
 6 -0.55883  -0.98999    0,785608 0.454472 -0.30874
 7  1.313973 -0.93645   -0.75936  0.748348  0.366501
 8  1.978716 -0.65364    0.066318 0.655470  2.046862
 9  0.824236 -0.21709    0.708716 0.869768  2.191926
10 -1.08804   0.283662  -0.81482  0.130148 -1.48905
11 -1.99998   0.708669   0.279928 0.375249 -0.63613
12 -1.07314   0.960170   0.591785 0.688185  1.166996
13  0.840334  0.976587  -0.83935  0.062679  1.040247
14  1.981214  0.753902   0.468189 0.178575  3.381881
15  1.300575  0.346635   0.433157 0.704439  2.784807
16 -0.57580  -0.14550   -0.83667  0.937345 -0.62063
17 -1.92279  -0.60201    0.618636 0.697482 -1.20868
18 -1.50197  -0.91113    0.238548 0.663530 -1.51102
19  0.299754 -0.99717   -0.80635  0.725215 -0.77955
20  1.825890 -0.83907    0.727211 0.472762  2.186793
21  1.673311 -0.47553    0.022004 0.228899  1.448678
22 -0.01770   0.004425  -0.74397  0.746051 -0.01120
23 -1.69244   0.483304   0.796689 0.795880  0.383434
24 -1.81115   0.843853  -0.19664  0.782893 -0.38105
```

```
25 -0.26470  0.997798 -0.64356  0.748695  0.838228
26  1.525116  0.907446  0.832941  0.245416  3.510921
27  1.912751  0.594920 -0.39697  0.575570  2.686266
28  0.541811  0.136737 -0.50138  0.506222  0.683391
29 -1.32726  -0.35492  0.840937  0.191429 -0.64982
30 -1.97606  -0.75968 -0.56349  0.349525 -2.94971
31 -0.80807  -0.97845 -0.31991  0.671893 -1.43454
32  1.102853 -0.95765  0.822002  0.585499  1.552695
33  1.999823 -0.70239 -0.68878  0.873207  1.481853
34  1.058165 -0.27516 -0.10998  0.084085  0.757103
35 -0.85636   0.219439  0.773141  0.551273  0.687489
36 -1.98355   0.660316 -0.77325  0.079198 -2.01729
37 -1.28707   0.939524  0.110371  0.035327 -0.20185
38  0.592737  0.988704  0.688595  0.399945  2.669983
39  1.927590  0.795814 -0.82206  0.508859  2.410204
40  1.490226  0.408082  0.320265  0.114586  2.333159
41 -0.31724  -0.07956  0.563231  0.314674  0.481096
42 -1.83304  -0.54772 -0.84094  0.987978 -2.23374
43 -1.66354  -0.88179  0.501669  0.744876 -1.29879
44  0.035403 -0.99996  0.396647  0.716758  0.148848
45  1.701807 -0.87330 -0.83290  0.312753  0.308351
46  1.803576 -0.53283  0.643776  0.180878  2.095398
47  0.247146 -0.06190  0.196265  0.844306  1.225812
48 -1.53650   0.424179 -0.79659  0.498035 -1.41089
49 -1.90750   0.806409  0.744121  0.685224  0.328250
50 -0.52474   0.991202 -0.02239  0.551240  0.995297
51  1.340458  0.933315 -0.72705  0.206753  1.753471
52  1.973255  0.646919  0.806435  0.442077  3.868687
53  0.791850  0.202135 -0.23891  0.423829  1.178896
54 -1.11757  -0.29213 -0.61840  0.065308 -1.96281
55 -1.99951  -0.71488  0.836707  0.223390 -1.65429
56 -1.04310  -0.96260 -0.43347  0.267109 -2.17207
57  0.872329 -0.97464 -0.46788  0.252564 -0.31763
58  1.985745 -0.74805  0.839334  0.901766  2.978789
59  1.273476 -0.33831 -0.59203  0.247839  0.590965
60 -0.60962   0.154251 -0.27956  0.102613 -0.63232
61 -1.93223   0.609055  0.814758  0.309997 -0.19842
62 -1.47836   0.914742 -0.70888  0.294933 -0.97757
63  0.334711  0.996467 -0.06592  0.657881  1.923132
64  1.840052  0.834223  0.759234  0.463877  3.897386
```

The following analyses will be done using SYSTAT. In the chapter on prediction analysis we outline several options to convert raw-data files into system files. The data must be available in ASCII code. Our present data are now in file FOUR.DAT, where the suffix .DAT is needed to enable SYSTAT to read it. The following commands, performed within the SYSTAT DATA module, transform the raw data into a system file:

SAVE FOUR (Initiates saving of data in system file FOUR.SYS.)

INPUT T, $V1$, $V2$, $V3$, RANDOM, CURVE (Names the variables in the file. Variable T indexes the observation points with $t = 1, ..., 64$; $V1$ to $V3$ denote the trigonometric functions described in formula 8; RANDOM is the random number variable. CURVE is the time series under analysis. It results from adding $V1$, $V2$, $V3$, and RANDOM for each value of t.)

GET FOUR (Reads ASCII code file FOUR.DAT.)

RUN (Starts saving of data.)

The command SWITCHTO SERIES transfers us to the SERIES module which reads the data from the system file and performs spectral analysis.

Step 2: Data Input

The SERIES module expects a series of numbers as input. The numbers must be stored on a system file. The following constraints apply:

1. The number of values must be equal to 2^x where x is an integer. If $t \neq 2^x$, the SYSTAT manual recommends adding zeros until the time series has 2^x values. If no zeros are added, the time series will be truncated to have 2^{x-1} values.

2. Although most programs do not explicitly specify this constraint, the observation points t must be equidistant. If the intervals between the observations differ, one can just number them. However, the estimated parameters will then generalize only to samples that follow the same spacing.

3. There must be no missing data. Since listwise or casewise deletion procedures are not applicable (there is only one series under analysis), missing data points must be estimated before analysis.

Step 3: Running the Program

Before performing spectral analysis, or the Fourier transform, the user must make a series of decisions. These decisions concern the transformations applied to the raw data before spectral analysis and transformations done before depicting the periodogram. Transformations of the raw data include removal of the mean, taking the logarithm of the raw data, squaring them, and detrending them. Detrending can be done by either removing the linear trend using the TREND variable name command or the DIFFERENCE variable name command. The second command substitutes differences between (adjacent) values for the original values, thus shortening the time series by one value. Without additional specification, the DIFFERENCE command calculates differences between adjacent values. The LAG specification enables the researcher to calculate differences between values further apart. For instance, the command

DIFFERENCE SEASON/LAG = 12

subtracts the first point from the twelfth, the second from the thirteenth, and so on. Other transformations are available. The transformed data can be saved.

If the researcher wishes to plot the data before decomposing them, he/she can use the PLOT command. Options include plotting of the raw data, plotting of standardized values, plotting the first X values, plotting of raw data transformed to lie in a specified interval, and plotting with specified labels.

For the present example, we choose to plot the raw data unchanged. The resulting printout is not reproduced here. The PLOT command will be illustrated as a tool to depict the periodogram.

The calculation of the periodogram is the central step in spectral analysis. The SERIES program uses the fast Fourier transform. Programs using this technique typically factor the trigonometric series and perform calculations using complex numbers. Such programs are much faster than standard programs, particularly if they decompose T, the number of observations into factors having the form 2^m. This is the reason for the first constraint mentioned above. The periodogram of a time series is generated using the FOURIER command. For our example we put

FOURIER CURVE/LAG = 32.

This command leads to the Fourier decomposition of the time series CURVES. The output, given in tabular form, contains the index, or the number of the assumed underlying frequencies. The largest frequency that can be uncovered has $f = 1/2$ because one needs at least two observations per period to be able to depict an oscillation. The LAG = 32 specification in the FOURIER command requests 32 frequencies in the printout, covering frequencies from 0 to 0.48. The default value is LAG = 15. However, since we generated a time series from series differing in cycle length, we needed a broader coverage than from 0 to 0.219.

Next to the index the program prints the indexed frequency. The frequencies are calculated at i/T with $i = 0, 1, \ldots$ The following two numbers are the real and the imaginary values of the calculated function. They are followed by the magnitude that is the maximum value that the function assumes for a given frequency. Similar to the function given in (3), SERIES calculates parameters for

$$y_t = a_1 + \sum_i (a_i \sin(jt) + a_{i+1} \cos(jt)) \qquad \text{for } i = 1, 3, 5, \ldots \qquad (9)$$
$$\text{and } j = 0, 2, 4, \ldots$$

The following two values give us the phase and periodogram values. High periodogram values suggest that the intensity of the wave is great for the indexed frequency. The following printout was generated using the following commands:

USE FOUR (Reads data from system file FOUR.SYS.)

OUTPUT@ (Sends output to both screen and printer.)

FOURIER CURVE/LAG = 32 (Generates spectral analysis of variable CURVE; prints results for 32 cycles.)

FOURIER COMPONENTS OF   CURVE

INDEX	FREQUENCY	REAL	IMAGINARY	MAGNITUDE	PHASE	PERIODOGRAM
1	0.00000	0.519	0.000	0.519	0.000	34.485
2	0.01563	0.033	-0.038	0.050	-0.860	0.323
3	0.03125	0.012	-0.016	0.020	-0.940	0.053
4	0.04688	0.102	-0.006	0.102	-0.055	1.337
5	0.06250	0.085	0.037	0.092	0.406	1.093
6	0.07813	0.365	0.331	0.493	0.736	31.088
7	0.09375	0.021	-0.029	0.036	-0.963	0.165
8	0.10938	0.016	-0.013	0.020	-0.692	0.053
9	0.12500	0.148	-0.004	0.148	-0.026	2.801
10	0.14063	0.169	-0.023	0.171	-0.133	3.724
11	0.15625	0.926	0.012	0.926	0.013	109.851
12	0.17188	-0.216	-0.011	0.216	-3.090	5.994
13	0.18750	-0.086	0.018	0.088	2.939	0.981
14	0.20313	-0.127	-0.012	0.127	-3.049	2.070
15	0.21875	-0.066	-0.024	0.070	-2.793	0.632
16	0.23438	-0.063	-0.003	0.063	-3.097	0.509
17	0.25000	-0.076	0.025	0.080	2.825	0.815
18	0.26563	0.006	0.018	0.019	1.261	0.047
19	0.28125	-0.021	0.013	0.025	2.579	0.082
20	0.29688	-0.018	-0.016	0.024	-2.433	0.073
21	0.31250	-0.042	0.029	0.051	2.531	0.334
22	0.32813	-0.038	0.046	0.060	2.261	0.463
23	0.34375	-0.320	0.318	0.451	2.359	25.994
24	0.35938	-0.015	-0.014	0.021	-2.393	0.055
25	0.37500	0.037	-0.039	0.054	-0.816	0.369
26	0.39063	0.015	-0.052	0.054	-1.289	0.372
27	0.40625	0.005	-0.013	0.014	-1.184	0.024
28	0.42188	-0.012	-0.038	0.040	-1.882	0.205
29	0.43750	0.017	-0.011	0.020	-0.566	0.050
30	0.45313	-0.018	0.009	0.020	2.663	0.053
31	0.46875	0.010	0.001	0.010	0.096	0.013
32	0.48438	-0.024	-0.003	0.025	-3.016	0.078

The periodogram displays four values larger than all others. The first is for frequency 0, the second is for frequency 0.078, the third is for frequency 0.156, and the fourth for frequency 0.343. The following plot depicts the periodogram. It was generated using the command PLOT CURVE/LAG = 32

```
PLOT OF     CURVE
NUMBER OF CASES =    64
MEAN OF SERIES =        0.120
STANDARD DEVIATION OF SERIES =       0.190

SEQUENCE PLOT OF SERIES

CASE       VALUE     0.009                                                    0.926
                     éáâáäéáâáäéáâáäéáâáäéáâáäéáâáäéáâáäéáâáäéáâáäéáâáäé
    1      0.519     ÜÜÜÜÜÜÜÜÜÜÜÜÜÜÜÜÜÜÜÜÜÜÜÜÜ
    2      0.050     ÜÜÜ
    3      0.020     Ü
    4      0.102     ÜÜÜÜÜÜ
    5      0.092     ÜÜÜÜÜ
    6      0.493     ÜÜÜÜÜÜÜÜÜÜÜÜÜÜÜÜÜÜÜÜÜÜÜÜ
    7      0.036     ÜÜ
    8      0.020     Ü
    9      0.148     ÜÜÜÜÜÜÜÜ
   10      0.171     ÜÜÜÜÜÜÜÜÜ
   11      0.926     ÜÜÜÜÜÜÜÜÜÜÜÜÜÜÜÜÜÜÜÜÜÜÜÜÜÜÜÜÜÜÜÜÜÜÜÜÜÜÜÜÜÜÜÜÜÜÜÜÜ
```

```
  12      0.216      ʊʊʊʊʊ̈ʊ̈ʊ̈ʊ̈ʊ̈ʊ̈
  13      0.088      ʊ̈ʊ̈ʊ̈ʊ̈ʊ̈
  14      0.127      ʊ̈ʊ̈ʊ̈ʊ̈ʊ̈ʊ̈
  15      0.070      ʊ̈ʊ̈ʊ̈ʊ̈
  16      0.063      ʊ̈ʊ̈ʊ̈
  17      0.080      ʊ̈ʊ̈ʊ̈ʊ̈
  18      0.019      ʊ̈
  19      0.025      ʊ̈
  20      0.024      ʊ̈
  21      0.051      ʊ̈ʊ̈ʊ̈
  22      0.060      ʊ̈ʊ̈ʊ̈
  23      0.451      ʊ̈ʊ̈ʊ̈ʊ̈ʊ̈ʊ̈ʊ̈ʊ̈ʊ̈ʊ̈ʊ̈ʊ̈ʊ̈ʊ̈ʊ̈ʊ̈ʊ̈ʊ̈ʊ̈ʊ̈
  24      0.021      ʊ̈
  25      0.054      ʊ̈ʊ̈ʊ̈
  26      0.054      ʊ̈ʊ̈ʊ̈
  27      0.014      ʊ̈
  28      0.040      ʊ̈ʊ̈
  29      0.020      ʊ̈
  30      0.020      ʊ̈
  31      0.010      ʊ̈
  32      0.025      ʊ̈
SERIES IS TRANSFORMED
```

(Notice that SERIES plots magnitude or VALUE against frequency; *cf.*, the above protocol of the Fourier decomposition.)

The first large value of the periodogram indicates that the zero frequency or the trend explains a great portion of the variance of the time series. This is somewhat unexpected, considering the way we constructed our data. However, a closer look at the first figure suggests that our time series may not be trendless. The next three peaks of the depicted function suggest a short frequency ($f = 0.078$), a medium frequency ($f = 0.156$), and a larger frequency ($f = 0.343$) component. A second medium frequency ($f = 0.171$) might be considered. The three largest values for frequency correspond with our intentions when we generated our data. The fourth largest value could be an artifact, resulting from adding the random numbers.

To evaluate the peaks statistically, we apply the significance test given in (7). For frequency 0.0 or index 1, we obtain $F(2,63) = 11.45$, for which $p<0.001$. For frequency 0.08 or index 6, we obtain $F(2,62) = 10.14$ with $p<0.001$; for frequency 0.16, or index 11, we obtain $F(2,62) = 60.53$ with $p<0.001$; and for frequency 0.34, or index 23, we obtain $F(2,62) = 8.263$ with $p<0.001$. The next largest value in the periodogram is I = 5.994 for frequency 0.17, or index 12. This peak is not significant ($F(2.62) = 1.731$, $p = 0.1856$).

We conclude from these results that the spectral analysis detected all three of the trigonometric functions we had used to generate the time series. The significant trend is somewhat unexpected. We have to assume the trend is a result of both the way the time series was calculated and the random numbers (see below). To illustrate spectral analysis for a demeaned time series, we reanalyze the same data, but we eliminate the mean before estimating the periodogram. The following commands generate a detrended spectral analysis of variable CURVE.

USE FOUR (Reads data from system file FOUR.SYS.)

MEAN CURVE (Removes the arithmetic mean from the time series of variable CURVE.)

FOURIER CURVE/LAG = 32 (Generates spectral analysis of CURVE.)

The resulting periodogram follows below.

FOURIER COMPONENTS OF   CURVE

INDEX	FREQUENCY	REAL	IMAGINARY	MAGNITUDE	PHASE	PERIODOGRAM
1	0.00000	0.000	0.000	0.000	0.000	0.078
2	0.01563	0.033	-0.038	0.050	-0.860	0.323
3	0.03125	0.012	-0.016	0.020	-0.940	0.053
4	0.04688	0.102	-0.006	0.102	-0.055	1.337
5	0.06250	0.085	0.037	0.092	0.406	1.093
6	0.07813	0.365	0.331	0.493	0.736	31.088
7	0.09375	0.021	-0.029	0.036	-0.963	0.165
8	0.10938	0.016	-0.013	0.020	-0.692	0.053
9	0.12500	0.148	-0.004	0.148	-0.026	2.801
10	0.14063	0.169	-0.023	0.171	-0.133	3.724
11	0.15625	0.926	0.012	0.926	0.013	109.851
12	0.17188	-0.216	-0.011	0.216	-3.090	5.994
13	0.18750	-0.086	0.018	0.088	2.939	0.981
14	0.20313	-0.127	-0.012	0.127	-3.049	2.070
15	0.21875	-0.066	-0.024	0.070	-2.793	0.632
16	0.23438	-0.063	-0.003	0.063	-3.097	0.509
17	0.25000	-0.076	0.025	0.080	2.825	0.815
18	0.26563	0.006	0.018	0.019	1.261	0.047
19	0.28125	-0.021	0.013	0.025	2.579	0.082
20	0.29688	-0.018	-0.016	0.024	-2.433	0.073
21	0.31250	-0.042	0.029	0.051	2.531	0.334
22	0.32813	-0.038	0.046	0.060	2.261	0.463
23	0.34375	-0.320	0.318	0.451	2.359	25.994
24	0.35938	-0.015	-0.014	0.021	-2.393	0.055
25	0.37500	0.037	-0.039	0.054	-0.816	0.369
26	0.39063	0.015	-0.052	0.054	-1.289	0.372
27	0.40625	0.005	-0.013	0.014	-1.184	0.024
28	0.42188	-0.012	-0.038	0.040	-1.882	0.205
29	0.43750	0.017	-0.011	0.020	-0.566	0.050
30	0.45313	-0.018	0.009	0.020	2.663	0.053
31	0.46875	0.010	0.001	0.010	0.096	0.013
32	0.48438	-0.024	-0.003	0.025	-3.016	0.078

The periodogram shows that the removal of the mean was very successful. The frequency 0.0 explains no portion of the variance. Again, there are the three expected frequencies largely contributing to the variance explanation. The significance tests for these peaks revealed only significant results (for frequency 0.08 $F(2,62) = 12.15$, for frequency 0.16 $F(2,62) = 85.29$, for frequency 0.34 $F(2,62) = 9.84$; for all Fs $p<0.001$). The explanatory value of frequency 0.17 is still not significant ($F(2,62) = 2.02$, $p = 0.1413$). Notice that the F values are different even though the periodogram values are still the same. The reason is that the sum in the denominator in (7) is reduced by the contribution of frequency 0.0. The following printout shows the magnitude by frequency plot for this result. We used the command

$$\text{PLOT CURVE/LAG} = 32, \text{ CENTER}$$

to illustrate what values deviate most dramatically from the mean of the periodogram.

```
PLOT OF     CURVE
NUMBER OF CASES =    64
MEAN OF SERIES =         0.112
STANDARD DEVIATION OF SERIES =      0.184

SEQUENCE PLOT OF SERIES

CASE       VALUE       0.000                                                 0.926
                          éáàáàéáàáàéáàáàéáàáàéáàáàéáàáàéáàáàéáàáàéáàáàéáàáàéáàáàé
            1       0.000     ÜÜÜÜÜÜ
            2       0.050     ÜÜÜÜÜ
            3       0.020     ÜÜÜÜÜ
            4       0.102     ÜÜ
            5       0.092     ÜÜÜ
            6       0.493             ÜÜÜÜÜÜÜÜÜÜÜÜÜÜÜÜÜÜ
            7       0.036     ÜÜÜÜÜÜ
            8       0.020     ÜÜÜÜÜÜ
            9       0.148     ÜÜ
           10       0.171     ÜÜÜÜ
           11       0.926     ÜÜÜÜÜÜÜÜÜÜÜÜÜÜÜÜÜÜÜÜÜÜÜÜÜÜÜÜÜÜÜÜÜÜÜÜÜÜÜÜ
           12       0.216     ÜÜÜÜÜÜ
           13       0.088     ÜÜÜ
           14       0.127     Ü
           15       0.070     ÜÜÜÜ
           16       0.063     ÜÜÜÜ
           17       0.080     ÜÜÜ
           18       0.019     ÜÜÜÜÜÜ
           19       0.025     ÜÜÜÜÜÜ
           20       0.024     ÜÜÜÜÜÜ
           21       0.051     ÜÜÜÜÜ
           22       0.060     ÜÜÜÜ
           23       0.451          ÜÜÜÜÜÜÜÜÜÜÜÜÜÜÜÜÜ
           24       0.021     ÜÜÜÜÜÜ
           25       0.054     ÜÜÜÜÜ
           26       0.054     ÜÜÜÜÜ
           27       0.014     ÜÜÜÜÜÜ
           28       0.040     ÜÜÜÜÜ
           29       0.020     ÜÜÜÜÜÜ
           30       0.020     ÜÜÜÜÜÜ
           31       0.010     ÜÜÜÜÜÜÜ
           32       0.025     ÜÜÜÜÜÜ
```

The following rules are often cited for application of spectral analysis to empirical data (*cf.* Metzler and Nickel, 1986).

1. The observation points must be equidistant.

2. The time series should have 100 or more observation points. Spectral analysis can be performed for shorter time series also ($50 \leq T \leq 100$) and the results will be trustworthy. However, as usual for small data sets, tests will show significant results only for those frequencies that explain a very large portion of the variance. Numerically, spectral analysis can be performed with very short time series; however, the shorter the time series, the more likely significance test will have no sufficient power.

3. Trends will appear as frequencies $f = 0$ for which the periodogram has a maximum. However, cycles of length $\geq T$ can result in such local maxima of $f = 0$. Trends (and seasonal effects) should, therefore, be eliminated before spectral analysis.

4. High frequencies can fuse with lower frequencies, a phenomenon called *aliasing*. This effect can lead to misinterpretations.

Programs for spectral analysis are included in most major program packages. Programs for PCs include SYSTATs SERIES that we used here and a program distributed by Brockwell, Davis, and Mandatino (1987). None of these PC programs includes the significance test used here.

Chapter 9   Time-Series Models

1. Summary of Method

Time series are generally thought of as any sequence of observations generated through time (Schmitz, 1990; Vandaele, 1983; Glass *et al.*, 1975). They
are marked by a characteristic dependency among data points, that is, a person's score at any one time is probably related to that person's score on the
same variable at any other time. A number of different strategies exist to describe this kind of data. Such methods include repeated measures ANOVA
(Bock, 1975; McCall and Appelbaum, 1973; Games, 1990), Tucker analysis
(Tucker, 1966), polynomial curve fitting, *P*-technique factor analysis (Jones
and Nesselroade, 1989), spectral analysis (Larsen 1990), and log-linear
modeling of longitudinal data (Upton, 1978). This chapter considers the
family of Box–Jenkins or ARIMA models (Box and Jenkins, 1976; Cook
and Campbell, 1979) that can be used to model either intraindividual
change (a time series represented repeatedly on an individual by a single
variable) or characteristics of summary statistics measured repeatedly on a
particular group.

    ARIMA (autoregressive integrated moving average) models attempt to
account for a time series by considering configurations of the series to be determined by the autocorrelations in the data along with a set of shocks to the

system. Consider the series pictured in Figure 1. This data could represent a time-dependent process to be modeled.

The strategy of a method such as polynomial curve fitting would attempt to find a mathematical function that described the curve of the data. This function takes the general form,

$$Y_t = f(t) \tag{1}$$

in which the value of time, t, is used as the independent variable in the computation. ARIMA, on the other hand, looks at the pattern of autocorrelation among occasions. This means that the predictor of Y is not t, but is instead the value of Y on prior occasions. This moves the analysis from the time domain into the frequency domain. Time is no longer a variable, but is now an index. The assumption that measures are equally spaced is now required, although covariance matrices can be computed that will account for unequal spacing (McArdle and Aber, 1990).

The correlations and covariances computed to identify the process are based on the number of points in the series. This requires a large enough N (the number of observations of the phenomenon) to allow a stable coefficient to be computed. Even with a relatively high N, the highest-order lagged

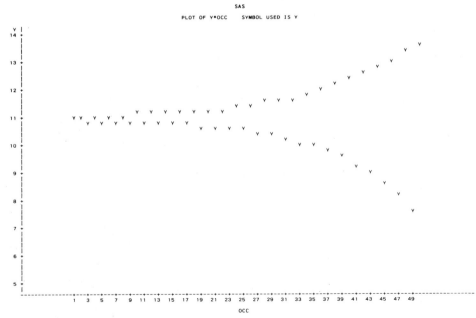

FIGURE 9.1. Representation of a time-dependent process with Y used as symbol in plot of Y*OCC (SAS).

covariances will be based on many fewer cases (i.e., the lag k covariances will be computed on N-k pairs of data).

Assuming that the series pictured is in some sense caused by an underlying process, one could consider these observations to be the realization of the process (Vandaele, 1983). Box–Jenkins models put time-series data through a set of filters that attempt to model the underlying process creating the pattern manifested in the data.

The first of these filters represents the requirement that a time series must be stationary to be interpretable. Stationarity includes two requirements for describing the series. A stable variance over time is necessary for the Box–Jenkins model. Since predictions made from the time series are based on prior values, increasing or decreasing variance could affect lagged covariances in some segment of the series. In Figure 1 variance increases over time.

Transformations can be used to stabilize the variance of a series (Vandaele, 1983). Three suggested are the log tranformation, the square-root transformation, and any of the family of power tranformations (Vandaele, 1983). When transforming data, it is important to plot the data both before and after the tranformation to see if the desired effect was achieved. The log tranformation in particular tends to overcompensate for the pattern of increased variation.

A second criterion for stationarity requires no overall trend (either up or down in the data). The models to be discussed attempt to describe the series in terms of the deviations from some typical expected or average value. With a general increasing or decreasing trend, no average based on the data can be considered typical. If there is no overall trend, each data point can be considered drawn from a distribution described by a single population mean. To detrend the data consecutive differencing is suggested. So if the time series can be described by the set of values

$$y_{series} = y_1, y_2, y_3, ..., y_n, \qquad (2)$$

the detrending can be accomplished be differencing. The above series can be differenced by

$$
\begin{aligned}
z_1 &= y_2 - y_1 \\
z_2 &= y_3 - y_2 \\
z_3 &= y_4 - y_3 \\
\cdot &\quad \cdot \quad \cdot \\
\cdot &\quad \cdot \quad \cdot \\
\cdot &\quad \cdot \quad \cdot \\
z_{n-1} &= y_n - y_{n-1}
\end{aligned}
\qquad (3)
$$

with the new series having one fewer observation. If an overall trend remains after this first differencing, the process can be repeated.

If the data tend to be seasonal (as in Figure 1), the series will manifest some kind of periodic change with the period representing the seasonal cycle. For example, weekly trends would cycle with a period of seven days; yearly trends would cycle with a period of twelve months. Any time series can be adjusted for these trends. In the case of data measured quarterly, a seasonal trend can be removed by

$$
\begin{aligned}
z_1 &= z_5 - z_1 \\
z_2 &= z_6 - z_2 \\
z_3 &= z_7 - z_3 \\
z_4 &= z_8 - z_4 \\
z_5 &= z_9 - z_5 \\
&\quad\vdots \\
z_{n-5} &= z_{n-1} - z_{n-5}.
\end{aligned}
\tag{4}
$$

As with consecutive differencing, seasonal differencing may be applied more than once if a single differencing does not remove the seasonal trend.

Once the model is considered stationary, it can be identified. The identification process first asks how many prior data points are necessary to make a good prediction of the current point. This number can be determined by looking at lagged autocorrelations. The first-order lag correlates each data point with its prior value. The second-order lag correlates each current value with the value with two occasions prior. This process continues with the third-, fourth-, *etc.*, lags until you run out of data points. The pattern of autocorrelation and partial autocorrelation can then be compared to patterns for known processes to determine the appropriate degree of the autoregressive process. When the number of necessary lags is selected, a regression equation based on the required autocovariances can be used to estimate a set of parameters. This equation can be expressed as

$$
z_t = \emptyset_1 z_{t-1} + \emptyset_2 z_{t-2} + \ldots + a_t
\tag{5}
$$

where $\emptyset_t$ represents the coefficient for an autoregressive process of order 1, and a_t represents a stochastic process.

If a_t is not an independent stochastic process (pure white noise), the value at any occasion may be affected by prior shocks in addition to the current shock. So just as lagged values of the observation can be used as predictors, lagged values of the noise process can be included in the model of the process. As an example, the equation

$$z_t = \emptyset_1 \, z_t + \emptyset_2 \, z_{t-2} + a_t - \theta_1 \, a_{t-1} \tag{6}$$

represents a second-order autoregressive process in which the most recent shock also contributes to the current value; θ_1 represents a regression-type coefficient linking the prior shock with the current value of the series.

As with the autoregressive process, these models tend to generate characteristic patterns of the autocorrelations represented in the autocorrelation (*acf*) and partial autocorrelation functions (pacf).

Once the model is identified, the $\emptyset_1, \emptyset_2, ..., \emptyset_{t-k}$ and $\theta_1, \theta_2, ..., \theta_{t-l}$ parameters can be estimated to yield the ARIMA estimation equation.

If multiple variables are considered, each can be thought of as having its own time series. If one of the variables is measured on something considered to be a dependent variable, the functional relationship between that and one or a set of predictors may be important. A model that would consider such a relationship is called a transfer-function model or a multiple-time series (Vandaele, 1983). One kind of transfer-function model is the intervention model. This model can be used to assess the influence of an intervention on what is considered to be a purely stochastic process. For this type of model the current value of the outcome of dependent variable is represented as a function of current and prior values of the predictor variable

$$Y_t = v_o \, X_t + v_1 \, X_{t-1} + ... + e_t \tag{7}$$

where the v_t represent regression-type coefficients linking current and prior values of the predictor with the current value of the criterion. Such models can be identified using the cross-covariances or correlations (Vandaele, 1983) between the series.

To illustrate the computation of the two most general types of time series problems, we include examples of a univariate ARIMA models and a transfer-function model.

2. Outline of the Computational Procedures

SAS PROC ARIMA is used to compute the ARIMA models to be discussed. This procedure divides the process into three steps: identification, estimation, and forecasting.

Identification is accomplished by using functions of the autocorrelations and autocovariances of the time series to compare the process against known patterns of autocovariation. Options for looking at cross-correlated variables allow one to consider possible transfer-function models. Patterns of cross-covariation can be used to identify these.

Estimation gives regression-like parameter estimates for the model that is selected. Although these two steps can appear in the same run, they can represent different stages of model development.

Forecasting of both existing data points and future data points represents the third step of the ARIMA process. Forecasting or prediction of existing data points is one indication of the utility of the model, suggesting how well one might trust future predictions.

3. Computational Steps for an AR(1) MA(1) Model

Step 1:   A Simulation of Time Series Data

To demonstrate an ARIMA process with a single lagged term ($p=1$) and a single moving average term ($q=1$), we simulated 50 data points based on the equation

$$Y_t = .8*Y_{t-1} - .4*A_{t-1} + A \qquad (8)$$

where A represents a stochastic process.

The SAS data step used to simulate the data was

```
OPTIONS NOCENTER LINESIZE=80 NODATE NONUMBER;
CMS FILEDEF OUTFILE DISK ARIMA11 DATA E1;
DATA ARIMA1;
   SEED=12323433;
   ARRAY A(I) A1-A50;
      DO OVER A;
         A=RANNOR(SEED);
            END;
ARRAY Y(I) Y1-Y50;
   I=1;
   Y1=0;
   FILE OUTFILE;
   PUT I Y1;
DO I=2 TO 50;
   I=I-1;
   YPAST1=Y;
   APAST1=A;
   I=I+1;
      Y= -.8*YPAST1 + A -.4*APAST1;
         PUT I Y;
      END;
```

Step 2:   Computing the Model

SAS PROC ARIMA was used to estimate the model. Since the process was known, we computed that model directly with the following SAS program:

```
OPTIONS NOCENTER LINESIZE=80 NONUMBER NODATE;
CMS FILEDEF SURVEY DISK ARIMA11 DATA E1;
DATA ARIMA11;
 INFILE SURVEY;
 INPUT OCC Y;
PROC ARIMA;
 IDENTIFY VAR=Y;
 ESTIMATE P=1 Q=1;
 FORECAST LEAD=10 BACK=10 ID=OCC OUT=TSRESULT;
PROC PLOT DATA=TSRESULT;
 PLOT Y*OCC='Y' FORECAST*OCC='F' L95*OCC='L' U95*OCC='U'/OVERLAY;
```

Y is identified as the time-series variable. The model to be estimated is an AR(1) MA(1) series. After the model is estimated values beyond those collected will be forecast. In fact the program will try to forecast the last 10 data points observed (BACK=10) and will then predict 10 new data points (LEAD=10).

The output of this model represents the steps of model identification, estimation, and forecasting.

After descriptive statistics are generated, the program begins the identification part of the process by printing the autocorrelations and auto-covariances:

```
SAS

ARIMA PROCEDURE

NAME OF VARIABLE       =        Y
MEAN OF WORKING SERIES= -0.078038
STANDARD DEVIATION     =  1.83281
NUMBER OF OBSERVATIONS=       50
AUTOCORRELATIONS

LAG COVARIANCE CORRELATION -1 9 8 7 6 5 4 3 2 1 0 1 2 3 4 5 6 7 8 9 1
 0   3.35918    1.00000   !                    !********************!
 1  -2.8389    -0.84512   !  ******************!                    !
 2   2.17062    0.64618   !                    !*************       !
 3  -1.72422   -0.51329   !         *********  !                    !
 4   1.32347    0.39399   !                    !********            !
 5  -0.95295   -0.28360   !              ******!                    !
 6   0.48368    0.14399   !                    !***                 !
 7  -0.184851  -0.05503   !                   *!                    !
 8  -0.106053  -0.03157   !                   *!                    !
 9   0.422648   0.12582   !                    !***                 !
10  -0.674245  -0.20072   !                ****!                    !
11   0.983782   0.29286   !                    !******              !
12  -1.10886   -0.33010   !               *****!                    !
13   1.04322    0.31056   !                    !******              !
14  -0.846179  -0.25190   !               *****!                    !
15   0.669026   0.19916   !                    !****                !
16  -0.621369  -0.18498   !                ****!                    !
17   0.539676   0.16066   !                    !***                 !
18  -0.473469  -0.14095   !                 ***!                    !
19   0.308213   0.09175   !                    !**                  !
20  -0.218694  -0.06510   !                   *!                    !
21   0.22201    0.06609   !                    !*                   !
22  -0.126452  -0.03764   !                   *!                    !
23   0.049761   0.01481   !                    !                    !
24  -0.0119861 -0.00357   !                    !                    !
                          !.' MARKS TWO STANDARD ERRORS
```

Each correlation ('***') can be referenced against a confidence interval of two standard errors (dotted lines). An AR(k) process typically shows a large correlation beginning at lag k. The patterns of correlations for lags greater than k tend to decay exponentially. This data fits that pattern well. An MA(k) process tends to show a spike at the lag k correlation and then quickly dies out. Because both processes in this data appear at lag 1, the MA process could be masked by the AR process.

Autocorrelation functions (acf) tend to die out slowly. As a result, they are often difficult to categorize, especially when more than one process is superimposed on the plot. Partial autocorrelation functions ($pacf$) can very often give a more direct indication of the process. If the process is truly AR(1), a pattern of lagged partial correlations will show a large value at lag 1 and values of 0 for other lags. The partial autocorrelations for this problem are

PARTIAL AUTOCORRELATIONS

```
  LAG  CORRELATION  -1 9 8 7 6 5 4 3 2 1 0 1 2 3 4 5 6 7 8 9 1
    1    -0.84512   |      *******************|            .            |
    2    -0.23810   |              .      *****|        .               |
    3    -0.14241   |              .      ***|         .                |
    4    -0.11624   |              .       **|         .                |
    5     0.00636   |              .         |         .                |
    6    -0.22048   |              .     ****|         .                |
    7    -0.15269   |              .      ***|         .                |
    8    -0.24266   |              .    *****|         .                |
    9    -0.02774   |              .       *|          .                |
   10    -0.13842   |              .      ***|         .                |
   11     0.11342   |              .         |**        .               |
   12     0.05991   |              .         |*         .               |
   13    -0.07629   |              .      **|          .                |
   14     0.29983   |              .         |**        .               |
   15     0.07396   |              .         |*         .               |
   16    -0.02517   |              .        *|         .                |
   17     0.06740   |              .         |*        .                |
   18    -0.03108   |              .        *|         .                |
   19    -0.11622   |              .       **|         .                |
   20    -0.17801   |              .      ****|         .                |
   21    -0.03910   |              .        *|         .                |
   22    -0.04376   |              .        *|         .                |
   23    -0.01824   |              .         |         .                |
   24    -0.07707   |              .      **|          .                |
```

The pattern shows one non-zero partial correlation. One could typically expect two spikes in the pacf if the process were AR(2). Characteristic patterns of many different ARMA processes are readily available (Vandaele, 1983; Glass *et al.*, 1975).

SAS next provides a check that the pattern of autocorrelations could have been generated by a truly stochastic (white noise) process. The *chi*-square tests indicate that a systematic process underlies this data.

AUTOCORRELATION CHECK FOR WHITE NOISE

TO LAG	CHI SQUARE	DF	PROB	AUTOCORRELATIONS					
6	89.73	6	0.000	-0.845	0.646	-0.513	0.394	-0.284	0.144
12	106.78	12	0.000	-0.055	-0.032	0.126	-0.201	0.293	-0.330
18	127.35	18	0.000	0.311	-0.252	0.199	-0.185	0.161	-0.141
24	128.97	24	0.000	0.092	-0.065	0.066	-0.038	0.015	-0.004

Next come the estimates of the parameters.

```
ARIMA: CONDITIONAL LEAST SQUARES ESTIMATION

                            APPROX.
PARAMETER      ESTIMATE      STD ERROR    T RATIO   LAG
MU            -0.0727105     0.0422895     -1.72     0
MA1,1          0.480255      0.145562       3.30     1
AR1,1         -0.711373      0.115797      -6.14     1

CONSTANT ESTIMATE  = -0.124435

VARIANCE  ESTIMATE =  0.926623
STD ERROR ESTIMATE =  0.962612
AIC               =    140.99*
SBC               =    146.726*
NUMBER OF RESIDUALS=       50
* DOES NOT INCLUDE LOG DETERMINANT

CORRELATIONS OF THE ESTIMATES

              MU     MA1,1    AR1,1
MU          1.000    0.044    0.015
MA1,1       0.044    1.000    0.460
AR1,1       0.015    0.460    1.000


AUTOCORRELATION CHECK OF RESIDUALS

TO    CHI                           AUTOCORRELATIONS
LAG  SQUARE DF   PROB
 6    6.00  3  0.112  0.060 -0.013 -0.141 -0.050 -0.210 -0.185
12   11.80  9  0.225 -0.056  0.047  0.144  0.113  0.198 -0.103
18   16.66 15  0.340  0.121 -0.037 -0.085 -0.122 -0.063 -0.146
24   18.53 21  0.615 -0.041  0.074  0.113  0.035 -0.002 -0.001


MODEL FOR VARIABLE Y
ESTIMATED MEAN        =-0.0727105

 AUTOREGRESSIVE FACTORS
FACTOR 1
1+0.711373B**(1)

 MOVING AVERAGE FACTORS
FACTOR 1
1-.480255B**(1)


MODEL FOR VARIABLE Y
ESTIMATED MEAN        =-0.0727105

 AUTOREGRESSIVE FACTORS
FACTOR 1
1+0.711373B**(1)

 MOVING AVERAGE FACTORS
FACTOR 1
1-.480255B**(1)
```

As can be seen, the estimates generated by the program are close to those of equation 7. The t-ratios (estimate/standard error) indicate that both the AR(1) and MA(1) parameters are necessary to the model. The check of the autocorrelations of the residuals indicates that there is not much else to estimate in this data.

Finally, based on the parameter estimates, some forecasting is done.

FORECASTS FOR VARIABLE Y

OBS	FORECAST	STD ERROR	LOWER 95%	UPPER 95%	ACTUAL	RESIDUAL

--------FORECAST BEGINS--------

OBS	FORECAST	STD ERROR	LOWER 95%	UPPER 95%	ACTUAL	RESIDUAL
41	-0.3930	0.9626	-2.2797	1.4937	-1.3187	-0.9257
42	0.1551	1.4975	-2.7798	3.0901	1.2702	1.1151
43	-0.2348	1.7054	-3.5772	3.1076	-0.7099	-0.4751
44	0.0426	1.8014	-3.4882	3.5734	0.9678	0.9252
45	-0.1547	1.8482	-3.7771	3.4676	-1.2955	-1.1407
46	-0.0144	1.8714	-3.6822	3.6534	2.6129	2.6273
47	-0.1142	1.8830	-3.8048	3.5764	-1.6066	-1.4924
48	-0.0432	1.8889	-3.7453	3.6589	-0.0620	-0.0188
49	-0.0937	1.8918	-3.8016	3.6142	-0.3430	-0.2493
50	-0.0578	1.8933	-3.7686	3.6531	-0.4226	-0.3648

The first 10 points predicted were the last 10 data points of the time series. The predictions can be compared against the observed points to give an indication of the utility of the model as a forecasting mechanism.

A PROC PLOT (Figure 2) was used to look at the data. The plot made use of the output data set from PROC ARIMA which included the actual data,

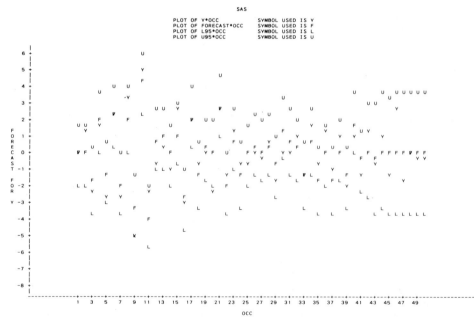

FIGURE 9.2. PROC PLOT of data set from PROC ARIMA. Plot of Y*OCC (symbol used is Y); plot of forecast*OCC (symbol used is F); plot of L95*OCC (symbol used is L); plot of U95*OCC (symbol used is U).

forecasted points, and an indication of the upper and lower confidence interval limits (*L*=lower 95%, *U*=upper 95%).

The conclusion from the output is that the process identified and estimated from the data represents an adequate description of the ARIMA process.

3. Computational Steps for an AR(2) MA(1) Model

Step 1: Another Data Simulation

Forty-eight data points were generated based on the following equation:

$$Y_t = .8*Y_{t-2} + .1*Y_{t-1} - .4*A_{t-1} + A. \tag{9}$$

The program used to generate this AR(2) MA(1) process was

```
OPTIONS NOCENTER LINESIZE=80 NONUMBER NODATE;
CMS FILEDEF OUTFILE DISK ARIMA31 DATA E1;

DATA ARIMA2;
  SEED=12323433;
  ARRAY A(I) A1-A50;
    DO OVER A;
        A=RANNOR(SEED);
            END;
ARRAY Y(I) Y1-Y50;
   I=1;
   Y1=0;
   FILE OUTFILE;
   I=2;
   Y2=.1*Y1+A2-.4*A1;
DO I=3 TO 50;
   I=I-2;
YPAST2=Y;
   I=I+1;
   YPAST1=Y;
   APAST1=A;
   I=I+1;
     Y= .1*YPAST1 - .8*YPAST2  -.4*APAST1 + A;
   END;
DO I=3 TO 50;
   Y=Y+400;
   WYE=Y;
   Y=SQRT(WYE);
   PUT I Y;
     END;
```

The original data produced by this equation tended to oscillate around the mean of the series with an amplitude that increased over time (see Figure 1). To remedy this violation of stationarity, we first created a positive series by adding a constant to the series. This was followed by a square-root transformation represented in the last DO loop. These two transformations produced a stationary series. A plot of the transformed data appears in Figure 3.

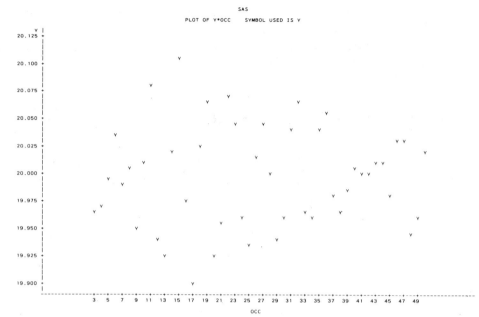

FIGURE 9.3.   Plot of transformed data with Y used as symbol in Y*OCC (SAS).

Step 2:   Computing the Model

The following SAS PROC ARIMA program was used to first identify and estimate the model and then to estimate 10 of the measured occasions and forecast 10 additional data points.

```
OPTIONS NOCENTER LINESIZE=80 NONUMBER NODATE;
CMS FILEDEF SURVEY DISK ARIMA31 DATA E1;
DATA ARIMA31;
  INFILE SURVEY;
  INPUT OCC Y;
PROC ARIMA;
  IDENTIFY VAR=Y;
  ESTIMATE P=2 Q=1 MAXIT=30;
  FORECAST LEAD=10 BACK=10;
PROC PLOT;
  PLOT Y*OCC='Y';
OPTIONS NOCENTER LINESIZE=80 NONUMBER NODATE;
```

The pattern of autocorrelations shows an oscillating series that decays exponentially.

```
SAS

ARIMA PROCEDURE

NAME OF VARIABLE      =        Y
MEAN OF WORKING SERIES=    19.9971
STANDARD DEVIATION    = 0.0455317
NUMBER OF OBSERVATIONS=        48
AUTOCORRELATIONS
```

```
LAG COVARIANCE CORRELATION -1 9 8 7 6 5 4 3 2 1 0 1 2 3 4 5 6 7 8 9 1
 0 0.00207314    1.00000    :                        :********************:
 1 .000027662    0.01334    :                        :.                   :
 2 -.00172804   -0.83354    :    ****************:                        :
 3 -.00013884   -0.06697    :                   .    *.                   :
 4 0.00138155    0.66640    :                   .    :*************       :
 5 .000204301    0.09855    :                   .    :**                  :
 6 -.00114502   -0.55231    :         ***********:                        :
 7 -.00029558   -0.14258    :                   ***:                      :
 8 .000839457    0.40492    :                   .    :*********           :
 9 .000336911    0.16251    :                   .    :***                 :
10 -.00054468   -0.26273    :                   .  *****:                  :
11 -.00032248   -0.15555    :                   .   ***:                   :
12 .000373798    0.18031    :                   .    :****                 :
13 .000414474    0.19993    :                   .    :****                 :
14 -.00032574   -0.15713    :                   .   ***:                   :
15 -.00046168   -0.22270    :                   .  ****:                   :
16 .000273365    0.13186    :                   .    :***                  :
17 .000469929    0.22668    :                   .    :*****                :
18 -.00025951   -0.12513    :                   .   ***:                   :
19 -.00050746   -0.24478    :                   .  *****:                  :
20 .000225445    0.10875    :                   .    :**                   :
21 .000487775    0.23528    :                   .    :*****                :
22 -.00017213   -0.08303    :                   .   **:                    :
23 -.00038969   -0.18797    :                   .  ****:                    :
24 .000146057    0.07045    :                   .    :*                    :
                            '.' MARKS TWO STANDARD ERRORS
```

The partial correlations that follow indicate that the process has at least a lag 2 relationship.

PARTIAL AUTOCORRELATIONS

```
       LAG CORRELATION -1 9 8 7 6 5 4 3 2 1 0 1 2 3 4 5 6 7 8 9 1
        1    0.01334    :                        .    :                    :
        2   -0.83387    :    ******************:                          :
        3   -0.11637    :                      .   **:                     :
        4   -0.09847    :                      .   **:                     :
        5   -0.02766    :                      .    *:                     :
        6   -0.09002    :                      .   **:                     :
        7   -0.11274    :                      .   **:                     :
        8   -0.21112    :                      . ****:                     :
        9   -0.09326    :                      .   **:                     :
       10    0.01246    :                      .    :                      :
       11   -0.00194    :                      .    :                      :
       12    0.04923    :                      .    :*                     :
       13    0.19696    :                      .    :****                  :
       14   -0.10206    :                      .  **:                      :
       15    0.00960    :                      .    :                      :
       16    0.03055    :                      .    :*                     :
       17    0.05460    :                      .    :*                     :
       18   -0.01182    :                      .    :                      :
       19   -0.03443    :                      .   *:                      :
       20   -0.06860    :                      .   *:                      :
       21   -0.11152    :                      .  **:                      :
       22   -0.04659    :                      .   *:                      :
       23    0.04705    :                      .    :*                     :
       24    0.01735    :                      .    :                      :
```

At this point patterns of autocorrelation and partial autocorrelation could be checked against those of known processes. We could, for example, first try a second-order only process, look at the residuals and decide if an additional term should be added. As we know in this case what the process is, we first check to make sure the pattern of autocorrelation is not trivial. The check of the autocorrelations for white noise yields

```
AUTOCORRELATION CHECK FOR WHITE NOISE

TO    CHI                        AUTOCORRELATIONS
LAG  SQUARE DF   PROB
  6   78.69  6  0.000   0.013 -0.834 -0.067  0.666  0.099 -0.552
 12   99.44 12  0.000  -0.143  0.405  0.163 -0.263 -0.156  0.180
 18  114.37 18  0.000   0.200 -0.157 -0.223  0.132  0.227 -0.125
 24  129.49 24  0.000  -0.245  0.109  0.235 -0.083 -0.188  0.070
```

This indicates correlations different than zero.
  The estimation of the process generates the following output

```
ARIMA: CONDITIONAL LEAST SQUARES ESTIMATION

                         APPROX.
PARAMETER    ESTIMATE    STD ERROR   T RATIO  LAG
MU           19.9976    0.00163681   9999.98    0
MA1,1         0.20277    0.176173       1.15    1
AR1,1         0.0915434  0.095991       0.95    1
AR1,2        -0.852867   0.082502     -10.34    2

CONSTANT ESTIMATE  =   35.2223

VARIANCE  ESTIMATE =.000641411
STD ERROR ESTIMATE = 0.0253261
AIC                =  -212.847*
SBC                =  -205.362*
NUMBER OF RESIDUALS=      48
* DOES NOT INCLUDE LOG DETERMINANT

CORRELATIONS OF THE ESTIMATES

             MU     MA1,1    AR1,1    AR1,2
MU         1.000    0.011    0.019    0.001
MA1,1      0.011    1.000    0.530    0.074
AR1,1      0.019    0.530    1.000   -0.006
AR1,2      0.001    0.074   -0.006    1.000


AUTOCORRELATION CHECK OF RESIDUALS

TO    CHI                        AUTOCORRELATIONS
LAG  SQUARE DF   PROB
  6    3.42  2  0.181   0.014 -0.056 -0.038 -0.093 -0.104 -0.191
 12    8.76  8  0.363  -0.007  0.028  0.055  0.164  0.167 -0.154
 18   10.03 14  0.760   0.080 -0.004 -0.054 -0.053 -0.026 -0.066
 24   11.31 20  0.938  -0.064  0.030  0.094  0.023  0.020 -0.002


MODEL FOR VARIABLE Y
ESTIMATED MEAN       =   19.9976

 AUTOREGRESSIVE FACTORS
FACTOR 1
1-.0915438**(1)+0.852867B**(2)

 MOVING AVERAGE FACTORS
FACTOR 1
1-0.20277B**(1)


MODEL FOR VARIABLE Y
ESTIMATED MEAN       =   19.9976

 AUTOREGRESSIVE FACTORS
FACTOR 1
1-.091543B**(1)+0.852867B**(2)

 MOVING AVERAGE FACTORS
FACTOR 1
1-0.20277B**(1)
```

FORECASTS FOR VARIABLE Y

OBS	FORECAST	STD ERROR	LOWER 95%	UPPER 95%	ACTUAL	RESIDUAL
-------FORECAST BEGINS-------						
39	20.0147	0.0253	19.9650	20.0643	19.9991	-0.0156
40	19.9921	0.0255	19.9422	20.0420	20.0000	0.0085
41	19.9826	0.0336	19.9168	20.0484	20.0084	0.0259
42	20.0009	0.0336	19.9351	20.0667	20.0104	0.0094
43	20.0107	0.0384	19.9354	20.0860	19.9812	-0.0295
44	19.9960	0.0384	19.9206	20.0713	20.0292	0.0332
45	19.9863	0.0416	19.9048	20.0677	20.0300	0.0437
46	19.9980	0.0416	19.9163	20.0796	19.9459	-0.0521
47	20.0073	0.0437	19.9217	20.0930	19.9607	-0.0466
48	19.9982	0.0438	19.9123	20.0841	20.0219	0.0237
				OCC		

The program reproduces the order-1 and order-2 autoregression parameters. It suggests, though, that the order-1 estimate may not be significantly different than 0. Having selected a small coefficient in the simulation, we did not consider this surprising.

The moving-average parameter was underestimated slightly. This term was also statistically not different than 0. A check of correlations among parameter estimates shows a relatively high correlation between the AR(1) and MA(1) parameter estimates. Removing one of these somewhat redundant terms might show a significant estimate for the remaining term. It is interesting to note that the simulation-equation coefficients are reproduced despite the square-root transformation.

The upshot of the estimation according to this output is that the process might be well-modeled by an order-2 autoregressive process and possibly either an order-1 moving-average process or an order-1 autoregressive. Knowing the simulation equation, we would vote for the moving-average process.

The residuals of the estimated values appear small. This suggests that the process can at least estimate the observed data. One test of the necessity of the AR(1) term would be to check the change in the residual values for the 10 estimated observations after that term was removed.

4. Computation of a Transfer Function Model

Step 1: Yet Another Data Simulation

To create variable, X, that was a lag 1 predictor of y (X_{t-1} predicts Y_t) we used the following program:

```
CMS FILEDEF OUTFILE DISK ARIMA41 DATA E1;
DATA ARIMA1;
   SEED=12323433;
   ARRAY A(I) A1-A50;
     DO OVER A;
        A=RANNOR(SEED);
          END;
```

```
ARRAY Y(I) Y1-Y50;
  I=1;
  Y1=0;
  FILE OUTFILE;
DO I=2 TO 50;
  I=I-1;
  YPAST1=Y;
  APAST1=A;
  I=I-1;
    Y= -.8*YPAST1 + A -.4*APAST1;
    END;
SEED2=111093;
ARRAY X(J) X1-X50;
  DO J=1 TO 49;
    I=J+1;
      X= .8*Y + .8*RANNOR(SEED2);
      END;
DO I=1 TO 49;
 J=I;
 PUT I Y X;
   END;
```

These data were generated by modifying the program used in the first simulation to create both X and the imagined criterion variable, Y. The SAS PROC ARIMA program used to analyze the data was

```
OPTIONS NOCENTER LINESIZE=80 NONUMBER NODATE;
CMS FILEDEF SURVEY DISK ARIMA41 DATA E1;
DATA ARIMA41;
  INFILE SURVEY;
  INPUT OCC Y X;
PROC ARIMA;
  IDENTIFY VAR=Y CROSSCORR=(X);
  ESTIMATE P=1 Q=1;
  FORECAST LEAD=10 BACK=10 ID=OCC OUT=TSRESULT;
PROC PLOT DATA=TSRESULT;
  PLOT Y*OCC='Y' FORECAST*OCC='F' L95*OCC='L' U95*OCC='U'/OVERLAY;
```

The program statements include the request to identify the series for Y which is expected to be cross-correlated with X.

The output first produced the information about the series, Y, that we saw above. In addition to that, the pattern of cross-correlations was shown to be

SAS

```
CORRELATION OF Y AND X
VARIANCE OF INPUT=   2.41183
NUMBER OF OBSERVATIONS=        49

CROSSCORRELATIONS

LAG COVARIANCE CORRELATION -1 9 8 7 6 5 4 3 2 1 0 1 2 3 4 5 6 7 8 9 1
-24  -0.273411   -0.09513   :                .     **:             .      :
-23   0.324292    0.11283   :                .      :**            .      :
-22  -0.343321   -0.11945   :                .     **:             .      :
-21   0.21337     0.07424   :                .      :*             .      :
-20  -0.0977622  -0.03401   :                .     *:              .      :
-19   0.0190912   0.00664   :                .      :              .      :
-18   0.0161344   0.00561   :                .      :              .      :
-17  -0.199786   -0.06951   :                .     *:              .      :
-16   0.527866    0.18366   :                .      :****          .      :
-15  -0.705634   -0.24550   :                .  ******:            .      :
-14   0.607703    0.21143   :                .      :****          .      :
-13  -0.73361    -0.25524   :                .  *****:             .      :
```

```
-12    0.886487    0.30843    ¦          ·        ¦******  ¦
-11   -1.01393    -0.35277    ¦       ·******·    ¦        ¦
-10    1.08458     0.37735    ¦          ·        ¦********¦
 -9   -0.832662   -0.28970    ¦       ·******·    ¦        ¦
 -8    0.636979    0.22162    ¦          ·      ¦****·     ¦
 -7   -0.376708   -0.13106    ¦          ·   ***¦          ¦
 -6    0.0670106   0.02331    ¦          ·      ¦·         ¦
 -5    0.243936    0.08487    ¦          ·      ¦**·       ¦
 -4   -0.780951   -0.27171    ¦       ·*****    ¦          ¦
 -3    1.23048     0.42811    ¦          ·      ¦********·*¦
 -2   -1.41051    -0.49074    ¦   ·**********   ¦          ¦
 -1    1.5863      0.55191    ¦          ·      ¦***********¦
  0   -2.00423    -0.69731    ¦·***************·¦          ¦
  1    2.55138     0.88768    ¦          ·      ¦********************¦
  2   -2.17953    -0.75830    ¦·****************¦          ¦
  3    1.59422     0.55466    ¦          ·      ¦***********¦
  4   -1.25903    -0.43804    ¦       ·*********¦          ¦
  5    0.965765    0.33601    ¦          ·      ¦*******·  ¦
  6   -0.612275   -0.21302    ¦          ·****  ¦          ¦
  7    0.286886    0.09981    ¦          ·      ¦**·       ¦
  8   -0.0937843  -0.03263    ¦          ·    *·¦          ¦
  9   -0.0548322  -0.01908    ¦          ·      ¦·         ¦
 10    0.209557    0.07291    ¦          ·      ¦*·        ¦
 11   -0.375838   -0.13076    ¦          ·   ***¦          ¦
 12    0.625183    0.21751    ¦          ·      ¦****·     ¦
 13   -0.758992   -0.26407    ¦       ·*****    ¦          ¦
 14    0.765752    0.26642    ¦          ·      ¦*****·    ¦
 15   -0.586586   -0.20409    ¦          ·****  ¦          ¦
 16    0.393023    0.13674    ¦          ·      ¦***·      ¦
 17   -0.311488   -0.10837    ¦          ·   **·¦          ¦
 18    0.29614     0.10303    ¦          ·      ¦**·       ¦
 19   -0.316774   -0.11021    ¦          ·   **·¦          ¦
 20    0.19257     0.06700    ¦          ·      ¦*·        ¦
 21   -0.174137   -0.06059    ¦          ·    *·¦          ¦
 22    0.21126     0.07350    ¦          ·      ¦*·        ¦
 23   -0.134092   -0.04665    ¦          ·    *·¦          ¦
 24   -.00973881  -0.00339    ¦          ·      ¦          ¦
                              ¦.' MARKS TWO STANDARD ERRORS
```

The pattern includes all possible lags between X and Y, including those negative lags in which a future value of X predicts a current value of Y (a leading indicator). The largest cross-correlation (.89) occurred at lag 1 with the lagged cross-correlations decreasing for greater lags. These, of course, are the results expected from the simulation. Regression-type estimates of the transfer-function parameters can be generated.

5. Discussion

We have presented three typical models. These can only be considered to represent a small taste of the many models subsumed under the general Box–Jenkins strategy. We have also somewhat indirectly suggested that the approach can be either confirmatory or exploratory. It is often the case that different models can do a good job of describing the same process (Vandaele, 1983). In that case, parameters used, as in all regression problems, should attempt to be parsimonious while at the same time including all substantively relevant parameters.

Developmental researchers should also remember that a single Box–Jenkins model can either look at the series for an individual or else look at the series based on the summary statistics of the group. It cannot do both. If the latter is done, those summary statistics are analyzed without reference to individual differences.

Chapter 10   Log-Linear Modeling

1. Summary of Method

Log-linear modeling (Bishop *et al.*, 1975) is a method for analysis of categorical data. When analyzing two-way cross- classifications, one can apply this method with three goals in mind (Goodman, 1984):

1. To analyze the joint distribution of the variables. Results of these analyses describe the distribution jointly displayed by the two cross-classified variables.

2. To analyze the association between two response variables. Results of these analyses depict the type and strength of the association between the two cross-classified variables.

3. To analyze the possible dependence of a response variable on an independent variable. Results of these analyses describe the conditional probabilities of the states of the response variable, given the states of the independent variable.

These goals apply accordingly to multiway tables.

Log-linear modeling typically proceeds in three steps. The first entails the specification of the assumptions to be tested. In other words, the researcher specifies a form of a joint frequency distribution, gives an assumed association pattern, or identifies dependent and independent variables. In the second step, the researcher estimates expected frequencies which exactly meet, under consideration of specified constraints, the specified assumptions. In the third step, the researcher applies goodness-of-fit tests to evaluate the closeness of the estimated expected frequencies to the observed frequencies and to evaluate the parameter estimates.

Considering, for instance, the three dichotomous variables A, B, and C, the general log-linear model allows one to estimate parameters for the following terms:

$$\log m = u_o + u_A + u_B + u_C + u_{AB} + u_{AC} + u_{BC} + u_{ABC} \tag{1}$$

where m denotes the observed cell frequencies. Parameter u_o is the "grand mean." It can be used to test the assumption that the cell frequencies follow a uniform distribution. The parameters u_A, u_B, and u_C concern the assumptions that the cell frequencies can be explained from the marginal proportions of the variables A, B, and C. The parameters with two subscripts can be used to test assumptions concerning pairwise interactions, and the last parameter refers to the triple interaction among all three variables. The model given in (1) has been termed the saturated model because it contains all possible terms. In most empirical research, one attempts to fit models more parsimonious than the saturated model.

The following two goodness-of-fit tests are most common in log-linear modeling: the Pearson X^2 and the likelihood ratio L^2. The test statistics are

$$X^2 = \sum_i (m_i - \hat{m}_i) / \hat{m}_i \tag{2}$$

and

$$L^2 = 2 \sum_i m_i \log(m_i/\hat{m}_i) \tag{3}$$

where i indexes the cells of the cross-tabulation, m_i denotes the observed cell frequencies and $\hat{m}_i$ the estimated expected cell frequencies. In (2) one typically uses the natural logarithm. If the specified model is true, X^2 and L^2 have the same asymptotic distribution. The decomposition characteristics of L^2 are better than those of X^2. If $m = 0$, Pearson's X^2 still can be calculated, and for L^2 one usually proceeds as if $\log 0 = 0$.

Most common are hierarchical log-linear models. In these models, interaction terms linking two or more variables imply the presence of all lower-order effects. For instance, the two-way interaction between the variables A and B implies the main effects of both A and B. Accordingly, the three-way interaction between the variables X, Y, and Z implies the two-way interactions XY, XZ, and YZ, and the main effects of X, Y, and Z.

Virtually all major statistical software packages contain programs that enable the researcher to apply hierarchical log-linear modeling. All of these programs provide an option for the researcher to specify the model to be fitted. Some modules, *e.g.*, HILOG in SPSSX (SPSS, 1983), provide an option to calculate all models until an *a priori* specified order of interaction has been reached.

In non-hierarchical log-linear modeling, interactions of any order can be specified without regress to lower-order terms. Examples of programs able to calculate non-hierarchical models include LOGLINEAR in SPSSX and the programs published by Haberman (1978) and Rindskopf (1987). Typically, these programs allow the user not only to specify non-hierarchical models, but also linear contrast-type comparisons between cells or groups of cells.

Common problems in log-linear modeling include the calculation of degrees of freedom, the handling of sparse tables, the analysis of weighted data, the specification of models for rates, and the interpretation of parameters. Clogg and Eliason (1987) discuss these problems and suggest solutions.

The following sections discuss two applications of log-linear modeling. The first is an analysis of the association pattern of three variables. A hierarchical model will be presented using the TABLES module in SYSTAT (Wilkinson, 1988). A non-hierarchical model will be presented using a PASCAL program by Rindskopf (1987). The second program can be used to analyze problems of the type discussed by Clogg, Eliason, and Grego (1990). In the second application we present the analysis of special effects. For this analysis, we also use Rindskopf's (1987) program.

2. Outline of the Computational Procedures for Hierarchical and Non-Hierarchical Log-Linear Models

This section outlines the computational steps for log-linear modeling on a microcomputer. The next section illustrates these steps in detail using a data example.

Log-linear modeling is done with categorical data. To fit a log-linear model, variables must be cross-tabulated. The SYSTAT module TABLES allows one to generate cross-tabulations from raw data available from a system file. Rindskopf's program expects interactive input of cell frequencies. If variables have to be categorized before cross- tabulation, SYSTAT's DATA module may be used. The models discussed in this paper are models for frequencies. (For models for rates see Clogg and Eliason, 1987; Rindskopf, 1987.)

SYSTAT's TABLES module reads data from SYSTAT system files. Interactive data input is possible in SYSTAT's DATA module. SYSTAT also provides an editor for data input or manipulation. Rindskopf's program prompts for the necessary information. In its present form it does not allow one to read data from files nor to form cross-tabulations from raw data. Both programs expect information on the dimensions of the cross-tabulation and the cell frequencies as input. This information allows one to do standard hierarchical log-linear modeling. For nonstandard testing of contrasts or dummy coding, Rindskopf's program prompts for vectors of the design matrix.

SYSTAT, release 4.0 (Wilkinson, 1988) can be used with IBM and compatible microcomputers under DOS 2.1 and higher. The resident memory must provide 512 K bytes in addition to the system. Other SYSTAT versions are available for APPLE, mini-, and main-frame computers. Rindskopf's program is available either as BASIC or PASCAL source code or compiled PASCAL programs for IBM PCs.

2.1   AN ANNOTATED DATA EXAMPLE: HIERARCHICAL LOG-LINEAR MODELING

The data we use for the following example were collected in an experiment on the effects of caffeine on memory performance (Lienert and Krauth, 1973). Participants were randomly assigned to four experimental groups. These groups resulted from crossing the two experimental variables, caffeine application (C) with levels 1 = caffeine administered and 2 = placebo, and type of instruction (S) with levels 1 = subjects were told they had received caffeine, and 2 = subjects were told they had received placebo. The response variable was memory performance. Immediately after being introduced to the experimental situation and receiving either a placebo or a dose of caffeine, subjects were presented with a series of pictures. Recall was tested immediately. A second recall followed after two weeks. Subjects were measured on whether memory performance (R) increased (+) or decreased (−). In the following sections, we illustrate how to analyze data of this kind with log-linear models.

2.1.1.   Hierarchical Log-Linear Modeling Using SYSTAT

Step 1:   Preparation of Data

There are three basic options for generating a SYSTAT system file. The first option involves keying in the data using SYSTAT's editor. The second option involves reading in an ASCII file written with a word processor and saving it as a system file. The DATA module provides this option. The third option involves keying in the data within the DATA module. The commands for the third option will be given here. (For more detailed descriptions see, for instance, the chapters on prediction analysis, scaling, Configural Frequency Analysis, or the SYSTAT manual.)

SAVE LOGEX (Initiates saving of data in system file LOGEX.SYS.)

INPUT C,S,R,F (Names the inputted variables.)

RUN (Starts input via keybord.)

These commands are followed by input of the cell indices and the cell frequencies. Numbers must be separated by commas or blanks. After completion of input, one keys in the command NEW which triggers the saving of the data and clears the workspace. The system file generated for the present example can be printed using the following commands in the DATA module:

USE LOGEX (Reads the system file LOGEX.SYS.)

OUTPUT@ (Sends output to printer.)

CASELIST (Initiates listing of contents of a system file.)

RUN (Starts the printing.)

The resulting printout is given below.

		C	S	R	F
CASE	1	1.000	1.000	1.000	19.000
CASE	2	1.000	1.000	2.000	48.000
CASE	3	1.000	2.000	1.000	24.000
CASE	4	1.000	2.000	2.000	27.000
CASE	5	2.000	1.000	1.000	12.000
CASE	6	2.000	1.000	2.000	56.000
CASE	7	2.000	2.000	1.000	13.000
CASE	8	2.000	2.000	2.000	39.000

Step 2: Data Input

The command SWITCHTO TABLES transfers the user from the DATA to the TABLES module. Since just printing did not send the data to the workspace, we have to read the data again. This is done, as in the DATA module, via USE LOGEX

The TABLES module expects a system file. After reading the data, the program presents the user with a list of variables.

Step 3: Running the Program

Before log-linear modeling, the user must cross-tabulate the variables under study. The following commands yield the cross-tabulation of the variables *C*, *S*, and *R*:

> WEIGHT = *F* (Variable *F*, which contains the cell frequencies, is interpreted as cell weights that function as cell frequencies. Notice that this command must be omitted when the data file contains raw data rather than cell indices and cell frequencies.

> OUTPUT@ (Sends output to printer.)

> TABULATE C*S*R (Generates cross-tabulation. The printout resulting from our present example appears below.)

```
TABLE OF        S       (ROWS) BY         R       (COLUMNS)
    FOR THE FOLLOWING VALUES:
                C      =     1.000

FREQUENCIES

            1.000      2.000     TOTAL
         ÖáááááááááááááááááááááC
   1.000 ·      19         48  ·      67
         ·                     ·
   2.000 ·      24         27  ·      51
         ááááááááááááááááááááááì
TOTAL          43          75         118

TABLE OF        S       (ROWS) BY         R       (COLUMNS)
    FOR THE FOLLOWING VALUES:
                C      =     2.000

FREQUENCIES

            1.000      2.000     TOTAL
         ÖáááááááááááááááááááááC
   1.000 ·      12         56  ·      68
         ·                     ·
   2.000 ·      13         39  ·      52
         ááááááááááááááááááááááì
TOTAL          25          95         120
```

To specify the log-linear model to be tested, one applies a format that is similar to the format of equation 1. In our example, the fully saturated model has the form MODEL C*S*R

Because the TABLES module calculates only hierarchical models, it is sufficient to specify the highest interaction term for each variable. This term includes all lower-order effects. For the present example, we test a model that makes the following assumptions:

1. After two weeks, fewer pictures will be remembered than immediately after presentation. Thus, variable *R* must have a main effect. In other words, when estimating expected frequencies we consider the marginal frequencies, of variable *R*.

2. Recall is associated with suggestion rather than caffeine. Thus we need a term for the association between S and R.

The second term contains the variables R and S. Since both variables' main effects are implied, the main effect for R does not need to be specified separately. The variable caffeine does not appear in the assumptions; therefore, it will not be included in the model specification. Consequently, the following model reflects our assumptions: MODEL S*R

For this model, we obtain the following goodness-of-fit results: $df = 4$, $X^2 = 7.65$, $p(X^2) = 0.105$, $L^2 = 7.72$, $p(L^2) = 0.102$. These tail probabilities suggest that the estimated expected frequencies do not differ significantly from the observed frequencies. Thus, we may work with this model.

SYSTAT offers a series of options for log-linear modeling. With these options the user can (1) manipulate the number of iteration steps performed, (2) specify the minimum difference between fitted values at two subsequent iterations that leads to the continuation of the iterations, and (3) add a constant to each cell frequency (Delta option). In addition, the user can consider structural zeros and save tables.

Main-frame packages give, in addition to global fit test statistics, chi-squared values for each effect. The next section explains how to specify and interpret these effects.

2.1.2. Hierarchical Log-Linear Modeling Using Rindskopf's Program.

To illustrate the use of Rindskopf's program for calculating hierarchical log-linear models, we use the same example as before.

Rindskopf's program expects the user to key in cell frequencies. In its present version, the program does not provide options for reading in files or forming cross-tabulations from raw data. The compiled PASCAL program is called Qualmod6. Thus we start the program by QUALMOD6

For this command we assume the computer is positioned at the drive that contains the program diskette. After starting the program, it responds by prompting the number of cells and the cell frequencies. The next prompt requires the user to specify whether the model is a model for rates or an ordinary log-linear model. In the present example we go with the latter. The following prompt concerns the design matrix. The program asks the user whether he/she wishes a standard design matrix or a custom-tailored one. In the former case the program generates the matrix, and results will be identical with those of other hierarchical programs. For the present purposes we go with this option. In the latter case the user must specify each vector of the design matrix. Non-hierarchical log-linear models can also be generated by deleting vectors from the design matrix.

In log-linear modeling, the design matrix has a function similar to the design matrix in the General Linear Model (*cf.* Clogg and Eliason, 1987; Clogg *et al.*, 1990; Langeheine, 1988; von Eye, *et al.*, in prep.). Any log-linear model can be expressed by

$$\log m = X^T \vec{b} \tag{4}$$

where m denotes the observed cell frequencies, X is the design matrix, and $\vec{b}$ is the parameter vector. In X we specify our assumptions on the observed frequency distribution in a way similar to the GLM. Examples appear below.

Degrees of freedom are calculated as follows. The maximum number of degrees of freedom is the number of cells, t, in a table minus 1. In addition, we subtract one degree of freedom for each of the vectors, c, in X. Thus, we obtain

$$df = t - c - 1. \tag{5}$$

There are two reasons why formula 5 may have to be corrected (Clogg and Eliason, 1987). The first is the existence of cells that have been blanked out and structural zeros. Each of these deprive the estimation process of degrees of freedom. The second reason is the existence of fitted zeros, or cells with estimated expected frequencies of zero. This type of cell also reduces the effective dimension of the cross-tabulation. Clogg and Eliason (1987) discuss solutions to these problems. When there are only a few instances, it is often sufficient to subtract one degree of freedom per structural or fitted zero.

For the present example we reproduce the hierarchical $S \times R$ model using Rindskopf's program. To do this, we first answer the following program prompts concerning the number of variables that have been crossed to form the contingency table and the number of categories of each variable. The next prompt leads to a model specification. The program asks for the highest interaction order to be included. For our three variables we specify first-order interactions because the second-order interaction would imply the saturated model.

The resulting design matrix contains vectors for the main effects of all three variables, and the first-order interactions $R \times S$, $R \times C$, and $S \times C$ (in this order). Since our model specification only implies the main effect vectors for R and S, and the $R \times S$ interaction, we must delete the third, fifth, and sixth vector of the design matrix. After viewing the results of the model with the six vectors, we answer the question as to whether we would like to run another model with yes and delete the rows, following the prompts. Recalculation leads to the following output.:

cell	observed freq	expected freq	standardized residual
1	19	15.50	0.889
2	48	52.00	-0.555
3	24	18.50	1.279
4	27	33.00	-1.044
5	12	15.50	-0.889
6	56	52.00	0.555
7	13	18.50	-1.279
8	39	33.00	1.044

Press any key to continue

goodness of fit tests
```
   likelihood ratio chi square =      7.725
   pearson chi square           =      7.648
   degrees of freedom           =      4
```

gamma	se	ga/se
-0.447	0.072	-6.171**
0.069	0.072	0.958
-0.158	0.072	-2.179**

Press any key to continue

Design matrix
```
  1   -1    1   -1    1   -1    1   -1
  1    1   -1   -1    1    1   -1   -1
  1   -1   -1    1    1   -1   -1    1
```

Corrections to gamma at last iteration (Should be 0 or close)

```
 0.00000   0.00000  -0.00000
```
Press any key to continue

The upper panel of the output displays the observed cell frequencies, the expected cell frequencies estimated under the hierarchical log-linear model $S \times R$, and the standardized residuals, which are defined as

$$r = (m - \hat{m})/\hat{m}^{1/2}. \tag{6}$$

In the present example we see that none of the standardized residuals is excessively large, i.e., greater than 1.96. The second panel of the output contains the results of the goodness-of-fit tests. As it should be, the test statistics are exactly the same as the ones calculated by SYSTAT. This is comforting.

The third panel gives the parameters of the model. Each of the parameters stands for one vector in X. The first in the present example is the parameter for the main effect of R. As can be seen from the printout, this parameter is statistically significant. We may therefore conclude that, after two weeks, recall rates are generally lower than they were immediately after presentation. The second parameter can be used to evaluate the main effect of S. This parameter is not significantly greater than zero. Therefore, we may not conclude that the suggestive information of a caffeine intake significantly affects recall rates. The third parameter concerns the $R \times S$ interaction. It suggests that the recall-rate differences vary across the levels of factor S.

The next panel of the output contains the design matrix. The vectors in this matrix are very useful for the interpretation of results. The first vector specifies that the numbers of individuals with increasing versus decreasing recall rates are compared. As can be seen from the matrix in the first panel, individuals with decreasing recall rates outnumber those with increasing rates under all experimental conditions. The second vector contrasts cell frequencies under the two suggestion conditions. This is a hypothesis that, from a substantive perspective, we are not interested in. The ratio 135:103 only tells us how many individuals participated under each condition. Statistically, this difference is not significant.

The third vector results from elementwise multiplication of the first two vectors. It specifies that, regardless of caffeine intake condition, the administration of caffeine intake increases the number of individuals who increase their recall rates relative to the number of individuals with decreasing recall rates. Also, suggestion of placebo intake leads to more individuals with decreasing recall rates after two weeks than individuals with increasing recall rates.

Since the second main effect is not significant, we might as well delete it and check if the resulting non-hierarchical model is still tenable. The deletion of the vector increases the number of degrees of freedom from 4 to 5 and the resulting goodness-of-fit tests are $X^2 = 8.64$ and $L^2 = 8.41$. None of these is significantly greater than admissible under alpha = 0.05. Thus, we may use the non-hierarchical model which, semantically, represents our assumptions better than the hierarchical model, because it only contains vectors that are relevant to our assumptions. More specifically, the main effect for S which was not part of our assumptions is part of the hierarchical model but not of the non-hierarchical model.

It should be noted that Rindskopf's program is not as user-friendly as the programs in the major software packages. The program does not provide an easy-to-use output option. The sample outputs were generated using the 'print screen' key. The next section illustrates the custom-tailored use of Rindskopf's program.

2.2 NON-HIERARCHICAL LOG-LINEAR MODELING WITH RINDSKOPF'S PROGRAM

The last section presented a first example of non-hierarchical log-linear modeling. The non-hierarchical model resulted from deleting vectors for lower-order terms from the design matrix and keeping the higher-order terms. The vectors for interactions resulted from standard elementwise multiplication of vectors for main-effects. This section illustrates the use of non-hierarchical models that contain vectors that do not result from multiplication of main effect vectors. We use the same example as in the last section.

We approach the caffeine–memory data under two hypotheses. The first hypothesis is, as before, that, after two weeks, reduced recall rates are more likely than increased recall rates. In the second hypothesis we assume that, regardless of caffeine intake, the number of individuals who display reduced recall rates is greater under the caffeine-suggestion condition than under the placebo-suggestion condition. This hypothesis is depicted by the following coding vector: 0, 1, 0, -1, 0, 1, 0, -1. Each 1 in this vector indicates our expectation of a cell frequency that is high relative to the frequencies in cells indexed by a -1. The following printout displays the results of this log-linear model.

cell	observed freq	expected freq	standardized residual
1	19	17.00	0.485
2	48	52.00	-0.555
3	24	17.00	1.698
4	27	33.00	-1.044
5	12	17.00	-1.213
6	56	52.00	0.555
7	13	17.00	-0.970
8	39	33.00	1.044

Press any key to continue

```
goodness of fit tests
    likelihood ratio chi square  =     8.255
    pearson chi square           =     8.327
    degrees of freedom           =     5
```

gamma	se	ga/se
-0.445	0.072	-6.161**
0.227	0.079	2.890**

Press any key to continue

```
Design matrix
  1  -1   1  -1   1  -1   1  -1
  0   1   0  -1   0   1   0  -1
```

Corrections to gamma at last iteration (Should be 0 or close)

```
0.00000  0.00000
Press any key to continue
```

The upper panel of the printout shows that the expected frequencies, estimated under this special model, are very close to the observed frequencies. None of the residuals is excessively big. Accordingly, the goodness-of-fit tests suggest that the model is tenable ($p(L^2 = 8.327) = 0.1392$, $df = 5$). Also, both parameters are significant, thus suggesting that our hypotheses cover a substantial portion of the variation in the table.

There are only a few programs that allow one to estimate non-hierarchical log-linear models. Examples include modules for log-linear modeling in BMDP and SPSSX. Of these, only the SPSSX program allows the user to specify values in the design matrix.

Chapter 11   Latent-Class Analysis

1. Summary of Method

Social science researchers often collect multivariate categorical data. These data can be analyzed from several perspectives. One can, for instance, focus on the relationships among observed variables. Examples of methods for analysis of interaction patterns in categorical variables include log-linear modeling and logistic regression. Other methods focus on underlying structures, dimensions, or classes. Examples of methods for analysis of dimensions in categorical variables include correspondence analysis and latent-class analysis (Clogg, 1981; Langeheine and Rost, 1988; Rindskopf, 1990).

Latent-class analysis (LCA) of categorical variables assumes one or more unobserved categorical latent variables with c categories or classes. The classes are mutually exclusive and exhaustive. In other words, each individual belongs to one and only one latent class, and all individuals belong to some class.

Parameters of LCA models are estimated so that the observed variables are independent of each other within each of the classes (axiom of local independence; see Clogg, 1988). Consider the three variables A, B, and C, observed in a sample of species. Suppose one of the assumed latent classes contains predators. For this and all other latent classes parameters are estimated so that A, B, and C are independent within each latent class.

If the resulting expected frequencies match the observed ones, the LCA model explains the relationship among the observed variables from a set of underlying classes. In this respect, LCA is analogous to factor analysis of continuous variables.

The probability of being in a latent class, or the unconditional latent-class probability, is denoted by Π_t^X where X is the latent variable and t is the latent class. The conditional probability $\Pi_{i\,t}^{AX}$ indicates how likely it is for an individual from class t to display category i of variable A. The value $\Pi_{i\,t}^{AX}$ indicates how strong the association between class t and category i is. This association is analogous to the factor loadings in factor analysis.

The axiom of local independence states that

$$\prod_{ij\ldots t}^{AB\ldots X} = \prod_{t}^{X} \prod_{it}^{AX} \prod_{jt}^{BX} \ldots \tag{1}$$

for each class in X. In words, the axiom of local independence states that the probability $\Pi_{i\,j\ldots t}^{AB\ldots X}$ within each class, t, can be estimated from the marginals of A, B, ... under the model of total independence of the variables A, B, ... Summing these probabilities over the classes yields the expected frequencies in the cross-tabulation under study.

To evaluate the model fit, one compares the expected frequencies with the observed ones. The test statistics to evaluate the fit are usually the same as in log-linear modeling. Specifically, one uses the Pearson or the likelihood ratio *chi*-square statistics. As in log-linear modeling, statistical tests are used to evaluate the closeness of the estimated expected to the observed frequency distribution. The null hypothesis states that the two distributions do not differ.

Recent developments of LCA concern the search for more general model formulations that allow one to derive latent-class and latent-structure models, the analysis of ordered classes, the development of LCA models for measuring, and the consideration of constraints (Langeheine and Rost, 1988). Applications include LCA of developmental models (Rindskopf, 1987, 1990) and LCA in adaptive testing (Kubinger, 1988).

2. Outline of the Computational Procedures

This section outlines the steps for computing LCA using a PC. The next section illustrates these steps in detail using an empirical data example. The program MLLSA by Clogg (1977) will be used.

MLLSA expects the user to input categorical data. A cross-tabulation can be generated using any of the major statistical software packages. The program expects cell frequencies as input. The data are part of an input file that also contains control information.

The MLLSA program reads data from an ASCII code file generated using a word processor. There is no option to key in data interactively. The program expects (1) information on the model to be estimated and (2) the cell frequencies. The information in the model, or the model specification, is given on a series of so-called option cards. Each line in the input file is called an "option card." Option cards contain both necessary and optional information. The following sections focus on the necessary information. For details on optional specifications, see the program manual (Clogg, 1977). We describe the necessary problem specifications in the required order.

The first card contains an 80-column title. The second card contains the number of manifest variables and the number of latent classes. Optional specifications on the second card concern the sample size, specifics of the iteration process, and output specifics. Also, one indicates on card 2 if one places model constraints on later cards.

On card 3, which is also necessary, one specifies the number of classes per latent variable. The format constrains the number of classes to be no greater than 99. This makes sense as it is very hard to imagine a social science application of LCA in which a latent variable has more than 99 classes.

Cards 4 and 5 are optional. Card 4 holds variable labels; card 5 holds value labels. Card 6 is necessary. It contains the data format given in FORTRAN terms. More specifically, one gives the data format as if the cell frequencies were real valued numbers. For instance if the maximum number of columns taken by a cell frequency is 5, and one has 10 frequencies on each data card, one writes (10f5.0). One card or 80 columns are available for the data format.

Card 7 contains the input data. The data must be given in the exact format specified on card 6. It is important to note that the order of the cell indices is such that the fastest varying variable is first, the second fastest varying the second, and so forth. For instance, if one analyzes three dichotomous variables, the order in which the cell frequencies must be placed on the data card is as follows: 111, 211, 121, 221, 112, 212, 122, 222.

The next card contains the start values for the latent class probabilities, the Π_t^x. Eight columns are available for each probability. Numbers must be given left-justified. Thus, the first probability is given in columns 1–8, beginning with the first. The second is given in columns 9–16, beginning with column 9, etc. If more cards than one are needed, additional cards may be added.

Card 9 contains the start values for the conditional probabilities in Formula 1. One begins with the variable that varies most rapidly. Here again, the input format expects the probabilities left-justified on eight columns each. For each variable one begins a new card. There is one (natural) constraint. The sum over all probabilities for each variable must be 1.0. Notice that the probabilities specified on cards 8 and 9 are used at the start of the iteration of the parameter estimation. In most instances, and unless explicitly specified as a model constraint, the final estimates will differ from the start values. Parameters may be fixed on cards 10 and 11.

Cards 10 and 11 are optional. They contain restrictions on the latent class parameters (card 10) and on conditional probabilities (card 11). Two types of constraint can be taken into account: (1) parameters can be fixed, and (2) parameters can be set equal to each other.

The cards described above contain all the information necessary for a complete run. The program will automatically start executing after reading the input information. During the run, the program provides one with the option to send the output to the printer or to an output file. In the latter case the user specifies path and name of the output file.

3.   An Annotated Data Example

The following data example, taken from Clogg (1988), analyzes $N = 1491$ respondents' answers to four questions concerning their satisfaction with family, friends, hobbies, and residence. The answers were scaled as follows: 1 = very great deal, a great deal; 2 = quite a bit, a fair amount, some, little, none. We analyze the cross-tabulation of answers using a two-class model. It is the goal of analysis to find out whether two classes (factors) can explain the relationships present in the observed frequency distribution. We use the PC version of Clogg's (1977) maximum-likelihood, latent-structure analysis program MLLSA.

Step 1:   Preparation of Data

The MLLSA program expects cell frequencies as input. The program does not provide a data management option. If data are not categorical or transformations are necessary to obtain certain categories, the data management modules in the major statistical software packages can be used (*e.g.*, SAS, SPSSX, BMDP, SYSTAT). The data for the present example are given in Clogg (1988, Table 1, p. 180).

Step 2: Data Input

The MLLSA program reads the model specification and the cell frequencies from an ASCII code file. Most word-processing programs allow one to generate ASCII code files. The input file must contain both the control-card information and the cell frequencies. The input file for the present example appears below.

```
C. C. Clogg's Satisfaction data (1988, p. 180)
0402                        1 1                 1
 2 2 2 2
(16f4.0)
 287  27  46  19 127  21  46  33 242  38  77  37 191  49 117 134
.8        .2
.8        .2        .2        .8
.8        .2        .2        .8
.8        .2        .2        .8
.8        .2        .2        .8
```

The first file of the input file contains the title of the run. The second line contains both necessary and optional information. The information on the number of observed variables (04) and latent classes (02) is necessary. The ones in columns 26, 28, and 42 are optional. The 1 in column 26 requests that the model fit for the independence model be included in the printout. This model can be considered a 1-class latent-class model. It is often used as a starting point for further analysis. The 1 in column 28 tells the program to assign respondents to latent classes. Solutions with only a few individuals correctly assigned are not trustworthy. The last 1 requests the output of standardized residuals which can be used for diagnostic purposes if the model does not fit.

The third line of the input file contains the number of categories for each of the observed variables. All variables in the example have two categories. The fourth line contains the input format. We have to input 16 cell frequencies, none of them requiring more than three columns. Thus, the format $(16F4.0)$ provides spaces between the cell frequencies and the data fit on one row. The fifth line contains the cell frequencies, right-justified, in the specified format. Each frequency occupies four columns.

The next lines contain the start values for the probabilities that the program estimates. First come the latent-class probabilities π_t^x. For each latent class one gives one probability. In this example we assume that there is one class of individuals satisfied with the majority of the areas addressed with the questions, and another class of people mostly dissatisfied. The probability statement indicates that we assume the first group is larger.

The following lines specify how likely it is for a respondent to fall into the "happy" group of largely satisfied individuals if he/she is largely satisfied versus less satisfied with his/her family. Specifically, we assume someone largely satisfied with his/her family will end up in the "happy" group with probability 0.8 and to end up in the "less happy" group with probability 0.2. Accordingly, someone less satisfied with his/her family will end up in the "happy" group with probability 0.2 and in the "less happy" group with probability 0.8. The following three lines give the same information for the variables friends, hobbies, and residence.

The resulting output appears below. We divided it into two sections. The first section repeats the input information. It is not given here. The second section contains the model of total independence, information on the iteration process, and the final result of the latent-class analysis.

```
C. C. Clogg's Satisfaction data (1988, p. 180)
TOTAL OF INPUT, AND EXPECTED    1491.      0.
    287.       27.       46.       19.      127.      21.       46.       33.      242.       38.

INPUT X'S   .8000000   .2000000

INPUT Y'S
 .8000000   .2000000   .2000000   .8000000
 .8000000   .2000000   .2000000   .8000000
 .8000000   .2000000   .2000000   .8000000
 .8000000   .2000000   .2000000   .8000000

MARGINALS FOR IND MODEL
1        1133.       358.
2         982.       509.
3         773.       718.
4         606.       885.

CHI-SQUARE FOR IND MODEL   LR=      363.217   PRSN=      486.612

X RESTRICTIONS

Y RESTRICTIONS

DEVIATION    .12806455E-02   ITERATIONS      20

DEVIATION    .47411043E-03   ITERATIONS      40

DEVIATION    .18092290E-03   ITERATIONS      60

DEVIATION    .69756203E-04   ITERATIONS      80

DEVIATION    .47729181E-04   ITERATIONS      88

FIT
    266.69      27.28      49.03      13.78     145.24      22.87      47.14      33.96
    265.42      36.08      71.55      43.16     165.36      53.05     122.56     127.81

CELL  OBSERVED  EXPECTED  (STDIZED RESID) (FREEMAN-TUKEY)
  1      287.    266.69  (  1.24 )        (  1.24 )
  2       27.     27.28  (  -.05 )        (  -.01 )
  3       46.     49.03  (  -.43 )        (  -.40 )
  4       19.     13.78  (  1.41 )        (  1.34 )
  5      127.    145.24  ( -1.51 )        ( -1.54 )
  6       21.     22.87  (  -.39 )        (  -.34 )
  7       46.     47.14  (  -.17 )        (  -.13 )
  8       33.     33.96  (  -.16 )        (  -.12 )
  9      242.    265.42  ( -1.44 )        ( -1.45 )
 10       38.     36.08  (   .32 )        (   .36 )
```

11	77.	71.55 (	.64)	(	.66)
12	37.	43.16 (	-.94)	(	-.93)
13	191.	165.36 (	1.99)	(	1.94)
14	49.	53.05 (	-.56)	(	-.53)
15	117.	122.56 (	-.50)	(	-.48)
16	134.	127.81 (	.55)	(	.56)

FINAL LR   14.291 PRSN   14.514 INDEX OF DISSIMILARITY   .0434

FINAL LATENT CLASS PROBABILITIES
.6656235   .3343765

FINAL CONDITIONAL PROBABILITIES

LATENT CLASS=		1	2	3	4
1.	1.		.9167	.4478	
1.	2.		.0833	.5522	
2.	1.		.8660	.2458	
2.	2.		.1340	.7542	
3.	1.		.6588	.2391	
3.	2.		.3412	.7609	
4.	1.		.5087	.2028	
4.	2.		.4913	.7972	

ASSIGNMENT OF RESPONDENTS TO LATENT CLASS J GIVEN MANIFEST RESPONSE I

CELL=	OBSERVED=	EXPECTED=	ASSIGN TO CLASS=	MODAL P=
1	287.	266.69	1	.9900
2	27.	27.28	1	.8798
3	46.	49.03	1	.8335
4	19.	13.78	2	.7304
5	127.	145.24	1	.9417
6	21.	22.87	1	.5436
7	46.	47.14	2	.5509
8	33.	33.96	2	.9433
9	242.	265.42	1	.9606
10	38.	36.08	1	.6425
11	77.	71.55	1	.5515
12	37.	43.16	2	.9169
13	191.	165.36	1	.7988
14	49.	53.05	2	.7737
15	117.	122.56	2	.8332
16	134.	127.81	2	.9855

PERCENT CORRECTLY ALLOCATED=   87.25 NUMBER CORRECTLY ALLOCATED=   1300.88   LAMB

The first part of the results contains information concerning the model of total independence of the answers to the questions on satisfaction with family, friends, hobbies, and residence. The program prints the marginals for this model, the likelihood ratio and the Pearson chi-squares. Both chi-squares are greater than the critical *chi*-square for $df = 13$. As expected, the model of independence is not tenable.

Right after the independence model the program prints the constraints on the latent-class model as imposed by the user. The printout shows that we did not put any constraints on the latent-class model. Information on the iteration process follows after the restrictions. The program prints the deviation score after every twentieth iterative step and after completion of the iterations. For the present example we needed 88 iterations.

The next part of the output contains the model evaluation. As requested, the program prints the standardized residuals $z = (o - e)/e^{\frac{1}{2}}$ and the Freeman–Tukey deviates $z = o^{\frac{1}{2}} + (o + 1)^{\frac{1}{2}} - (4e + 1)^{\frac{1}{2}}$. The overall likelihood ratio for this example is 14.291. For $df = 6$ this value has tail probability 0.0265 suggesting that the model is marginally acceptable. After the model evaluation follow the final latent-class probabilities. As expected, the "happy" class is larger than the "less happy" class.

The conditional probabilities printed next are to be read as follows: An individual from latent class 1 displays great satisfaction with his/her family with probability 0.917 and not so great satisfaction with probability 0.083. For individuals from latent class 2 these probabilities are 0.448 and 0.552, respectively. For the last set of conditional probabilities we notice that individuals from latent class 1 are largely satisfied with their residence with probability 0.509 and not quite so satisfied with probability 0.491. For individuals from the second latent class these probabilities are 0.203 and 0.797, respectively. Notice that the conditional probabilities for each latent class sum up to 1 because they are estimated such that they apply to every individual.

The probabilities that individuals belong to a certain latent class if they display a certain response pattern follow. They can be read as follows. If an individual shows the first response pattern, that is, pattern 1111, then he/she belongs to latent class 1 with probability .99. If an individual shows the second response pattern, 2111, then he/she belongs to latent class 1 with probability 0.88, and so forth. The present solution displays, for some patterns, assignment probabilities only marginally different from 0.5, which is chance assignment. Accordingly, only 87.25% of the sample are correctly allocated. The asymmetric similarity coefficient lambda (Goodman, Kruskal, 1954) between the observed responses and the latent classes is 0.619, indicating an overall gain of 61.87% achieved by the two-class model relative to the independence model.

To summarize, the present results suggest that the two-class model provides a marginally acceptable explanation of the observed frequency distribution. Other models and details concerning the parameters are discussed by Clogg (1988).

The MLLSA program is among the most widely distributed LCA programs in the social sciences. It can be requested from Scott R. Eliason, Population Issues Research Office, The Pennsylvania State University, 22 Burrowes Building, University Park, PA 16802.

Other programs include the Categorical Data Analysis System (CDAS) by Eliason (1988), which can be purchased from Dr. Eliason. Rindskopf (1990) explains how to use the IMSL routines for estimation of latent-class models. The IMSL FORTRAN routines are available in main-frame, mini-, and microcomputer versions.

Chapter 12

Analysis of Finite-Mixture Distributions

1. Summary of Method

In many applications of social science statistics researchers assume that samples are heterogeneous. To identify groupings of individuals methods of cluster analysis are used. Subjects within clusters are more similar to each other than to subjects from other clusters. Most methods of cluster analysis utilize concepts of similarity or distance in the cluster formation. In contrast, the analysis of finite-mixture distributions generates classifications from distributional assumptions (Everitt and Hand, 1981; Erdfelder, 1990).

Suppose d variables have been observed in a sample of N individuals. Then we have a d-dimensional random vector Y for each individual. The value y_{ij} contains the ith individual's measure on the jth variable, for $i = 1, ..., N$ and $j = 1, ..., D$. For the analysis of these data with finite-mixture distributions, we assume that the population we sampled the subjects from consists of c classes. These classes are disjunctive—that is, they do not overlap, and they are exhaustive—that is, they contain all members of the population. In addition, we assume that the observed frequency distribution is the mixture of c probability density functions.

Before analyzing the data, the researcher only assumes the number of

191

classes. Which individual belongs to which class is unknown. The goal of the analysis is to decompose the observed frequency distribution into c classes.

The *general finite mixture distribution model* is given by

$$f(Y = y) = \sum_{k=1}^{c} g(Y = y; \theta_k) p(X = x_k) \qquad (1)$$

(Everitt and Hand, 1981; Erdfelder, 1990). The term $f(Y = y)$ has been termed finite-mixture density function (Titterington *et al.*, 1985), and k indexes the unknown, or *latent classes* (*cf.* the chapter on latent-class analysis; Rindskopf, 1990). $p(X = x_k)$ is the probability that X assumes the value x_k, or the probability that an individual is a member of the kth class. θ_k is a parameter vector that may be different for each class. It contains the parameters of the density functions assumed to underlie the observed frequency distribution. Before analysis of data, explicit formulas must be inserted. Examples of distributions used for analysis of finite-mixture distributions include uni- and multivariate variants of normal, binomial, multinomial and Poisson density and probability functions.

A finite mixture has three sets of parameters:

1. the probability $p(X = x_k)$ that an individual falls into class k,
2. the values of the vector θ_k,
3. the number c of latent classes.

In the following applications of finite-mixture models, the researcher specifies the parameter c before analysis. Also, the researcher determines the probability density function of which the observed frequency distribution is assumed to be a mixture. All other parameters are estimated from the data. When analyzing empirical data, the researcher performs the following three steps (Erdfelder, 1990):

1. *Checking whether the model is identifiable.* A model is identifiable if there is a unique set of parameters for each latent class.

2. *Estimation of parameters.* If no constraints are placed, the number of parameters to be estimated is $c(d + 1) - 1$ where d denotes the number of elements in θ_k. The estimation itself is typically done using maximum-likelihood or estimation-maximization algorithms (EM; Dempster *et al.*, 1977). It is possible to place restrictions on the parameters. For instance, the researcher may assume that a class has a certain a priori probability.

3. *Testing the goodness-of-fit of a model.* Here the same methods are applied as in log-linear modeling (see chapter on log-linear modeling; Fienberg, 1980).

To be able to estimate the parameters the researcher must specify the type of density function assumed to underlie the observed frequency distribution. (Notice that this is not necessary in standard latent class analysis; see Rindskopf, 1990.) In the remainder of this chapter we give an example using the binomial distribution. In principle, however, any sampling distribution may be used. Examples include the univariate normal distribution, the multivariate normal distribution, the product multinomial distribution, and the binomial distribution that we use here. Standard latent-class analysis (see chapter on latent-class analysis) can be described as the analysis of mixtures of multinomial distributions (see Everitt and Dunn, 1988; Erdfelder, 1990).

The binomial distribution is the distribution of choice if the variables under study have a finite number of discrete categories. An example is the number of hits and misses in an examination. Suppose we analyze the univariate frequency distribution of variable Y. Y has m categories. The binomial distribution of Y is

$$g(Y = i; p_k) = \binom{n}{i} p_k^i q_k^{(n-i)} \tag{2}$$

where $i = 1, ..., m$ and $q_k = 1 - p_k$. Inserting of (2) into (1) yields the model of finite binomial mixtures (Erdfelder, 1990). These models have two sets of parameters: the probabilities of the latent classes and p_k, the success probability estimated for the members of the kth class.

2. Outline of the Computational Procedures

The following applications analyze mixtures of binomial distributions. Erdfelder's (1990) BINOMIX program will be used.

The following examples use categorical data; therefore, we assume the data either are already categorical or have been categorized before analysis. The BINOMIX program expects cell frequencies as input. It does not provide an option for generating frequency counts from raw data. Frequency counts and cross-tabulations can be generated using any of the major statistical software packages for PCs.

The BINOMIX program is fully interactive. Data are inputted via the keyboard. There is no option to read data from files. In a vignette that appears after the program starts, the program tells the user what input information it expects. This vignette appears below.

```
                            B  I  N  O  M  I  X
                ANALYSIS OF MIXTURES OF BINOMIAL DISTRIBUTIONS

-------------------------------------------------------------------------
USAGE :    BINOMIX COMPUTES MAXIMUM LIKELIHOOD ESTIMATORS OF THE
           PARAMETERS OF MIXTURES OF BINOMIAL DISTRIBUTIONS. M (THE
           NUMBER OF TRIALS PER SUBJECT) MUST BE A FIXED CONSTANT,
           BUT P (THE SUCCESS PROBABILITY) CAN VARY BETWEEN SUBJECTS.
           TWO CASES OF MIXTURES CAN BE ANALYSED:
           - P FOLLOWS A BETA DISTRIBUTION WITH PARAMETERS THETA1 AND
             THETA2 (BETA-BINOMIAL-MODEL);
           - P FOLLOWS A DISCRETE DISTRIBUTION (LATENT-CLASS-MODEL).
INPUT :    - JOB TITLE
           - THE NUMBER M OF TRIALS PER SUBJECT
           - THE NUMBER N(I) (0 <= I <= M) OF SUBJECTS WITH
             I SUCCESSES IN M TRIALS
           - KIND OF MODEL TO ANALYSE AND CONVERGENCE CRITERION
OUTPUT:    - FREQUENCY TABLE
           - PARAMETER ESTIMATES
           - RESULTS OF MODEL EVALUATION

           (PRESS E TO LEAVE PROGRAM OR ANY OTHER KEY TO CONTINUE)
```

After data entry, the program prompts the user to make a decision as to what statistical model he/she wishes to use. The choice is between the beta binomial and the latent-class model. After this choice, the user may specify a convergence criterion that is a threshold or a minimal value that must be reached for the program to terminate the iterations for the parameter estimation. Also, the user must specify the number of latent classes assumed to underlie the observed frequency distribution.

3. An Annotated Data Example

The following data example analyzes data from an investigation on the efficiency of a new drug (data adapted from Krause and Metzler, 1984). The drug was administered in a sample of 360 patients over a period of four weeks. The number of symptoms alleviated, m, was counted. The scale ran from no symptoms improved ($m = 0$) to all symptoms improved ($m = 5$). We analyze these data under two alternative assumptions. The first is there are two groups of patients. These groups may differ in size and in the success parameter, or their chances of getting cured by the drug. The second assumption states that there are three groups of patients, also differing in size and success parameter. BINOMIX will be used to generate a solution for both assumptions.

Step 1: Preparation of Data

BINOMIX expects cell frequencies as input. Data are keyed in. They only consist of an ordered string of frequencies. The order is determined by the order of the (ordinal) categories of the variable under study.

Step 2: Data Input

The program prompts for the information necessary to perform the decomposition of binomial mixtures. The first prompt concerns the "number of trials per subject." In the present example we consider each symptom a trial. As was explained earlier, the number of symptoms a patient can be freed of can range from 0 to 5. Next the program asks for the frequencies observed for each level of the dependent variable Y ("number of subjects with i successes"). We answer by keying in the cell frequencies. The following prompt gives the user the choice between a beta binomial model and a latent-class model. For this example we use the latent-class model (for the *beta* binomial model see Erdfelder, 1990).

The convergence criterion determines the precision of the iteration process. The smaller we set this value, the more iterations may be needed for a solution. For the present purposes we choose the value 0.0001 (default is 0.00000001). The data input is summarized in the following printout.

```
PROGRAM BINOMIX      DATE = 10-25-1989      TIME = 15:27:37      PAGE = 1
JOB TITLE : Workbook Data Example Finite Binomial Mixtures
------------------------------------------------------------------------

INPUT OF RAW DATA AND JOB CONTROL INFORMATION :

NUMBER OF TRIALS PER SUBJECT? 5

NUMBER OF SUBJECTS WITH 0 SUCCESSES? 91
NUMBER OF SUBJECTS WITH 1 SUCCESSES? 59
NUMBER OF SUBJECTS WITH 2 SUCCESSES? 62
NUMBER OF SUBJECTS WITH 3 SUCCESSES? 75
NUMBER OF SUBJECTS WITH 4 SUCCESSES? 44
NUMBER OF SUBJECTS WITH 5 SUCCESSES? 29

BETA-BINOMIAL-MODEL (=1) OR LATENT-CLASS-MODEL (=2)? 2

CONVERGENCE CRITERION (DEFAULT: 1D-08)? .0001

                    ( PRESS ANY KEY TO CONTINUE )
```

Step 3: Running the Program

After input of data and model specifications the program gives a histogram of the observed cell frequencies. This histogram appears below.

```
PROGRAM BINOMIX      DATE = 10-25-1989      TIME = 15:28:50      PAGE = 2
JOB TITLE : Workbook Data Example Finite Binomial Mixtures
------------------------------------------------------------------------

NUMBER OF TRIALS PER SUBJECT : 5

SUCCESSES  N OF SUBJECTS  % OF SUBJECTS I        H I S T O G R A M
------------------------------------------------------------------------
    0            91          25.2778   IXXXXXXXXXXXXXXXXXXXXXXXXXXXXXXXXXXX
    1            59          16.3889   IXXXXXXXXXXXXXXXXXXXXXXX
    2            62          17.2222   IXXXXXXXXXXXXXXXXXXXXXXXX
    3            75          20.8333   IXXXXXXXXXXXXXXXXXXXXXXXXXXXXXX
    4            44          12.2222   IXXXXXXXXXXXXXXXXX
    5            29           8.0556   IXXXXXXXXXXX
------------------------------------------------------------------------
TOTAL N OF SUBJECTS : 360

                    ( PRESS ANY KEY TO CONTINUE )
```

The key stroke to continue the analysis after presentation of the histogram triggers the iteration. If the specified convergence criterion is too small for a user's patience, the escape key stops the iteration and the program uses the latest parameter estimates for the following analyses. The first screen after the iterations shows the parameter estimates. The estimates for the present example are given below.

```
PROGRAM BINOMIX       DATE = 10-25-1989       TIME = 15:30:13       PAGE = 4
JOB TITLE : Workbook Data Example Finite Binomial Mixtures
------------------------------------------------------------------------------

RESULTS OF PARAMETER ESTIMATION FOR UNIVARIATE LATENT-CLASS-MODEL
WITH LOCAL BINOMIAL DISTRIBUTIONS:

    CLASS            P(CLASS)    P(SUCCESS / CLASS)
    ------------------------------------------------------
      1               0.58777        0.61951
      2               0.41223        0.09914
    ------------------------------------------------------

LAST MAXIMUM DEVIATION IN PARAMETER ESTIMATION : 8.885024833552024D-05
NUMBER OF ITERATIONS : 27

              ( PRESS ANY KEY TO CONTINUE )
```

The parameter estimates show that 41.32% of the population belong to class 2, and 58.68% to class 1. The probability for the drug to free patients of class 2 from symptoms is only 0.10. In the first class this probability is 0.62. The next screen displays the conditional probabilities for each group of patients and each number of cured symptoms. These probabilities are conditional on the success frequencies. The printout appears below.

```
PROGRAM BINOMIX       DATE = 10-25-1989       TIME = 15:31:20       PAGE = 5
JOB TITLE : Workbook Data Example Finite Binomial Mixtures
------------------------------------------------------------------------------

CONDITIONAL PROBABILITIES P(CLASS / SUCCESSES) :

SUCCESSES I  CLASS 1   CLASS 2
----------------------------------
    0     I  0.01882   0.98118
    1     I  0.22108   0.77892
    2     I  0.80772   0.19228
    3     I  0.98417   0.01583
    4     I  0.99891   0.00109
    5     I  0.99993   0.00007
----------------------------------

              ( PRESS ANY KEY TO CONTINUE )
```

The results show that patients freed of no or only one symptom most likely belong to class 2. Patients freed of two or more symptoms most likely belong to class 1. For patients freed of all five symptoms it is almost impossible to belong to class 2. We may conclude from these results that before application of this drug, differential diagnosis is most useful, because this drug is effective only in class 1 patients.

The last information of the program output concerns the model fit. For the present example we obtain the following printout.

```
PROGRAM BINOMIX       DATE = 10-25-1989       TIME = 15:32:08       PAGE = 6
JOB TITLE : Workbook Data Example Finite Binomial Mixtures
-----------------------------------------------------------------------------

EVALUATION OF LATENT-CLASS-MODEL WITH 2 LATENT CLASSES:

SUCCESSES   THEOR. PROB.   EXPECTED N   OBSERVED N   STD. RESID.
-----------------------------------------------------------------
    0          0.24927        89.74        91.00         0.13
    1          0.17274        62.19        59.00        -0.40
    2          0.15388        55.40        62.00         0.89
    3          0.20558        74.01        75.00         0.12
    4          0.16489        59.36        44.00        -1.99
    5          0.05364        19.31        29.00         2.20
-----------------------------------------------------------------

FIT INDICES:
PEARSON CHI-SQUARE            =    9.8178
LIKELIHOOD-RATIO CHI-SQUARE  =    9.5312

NEW JOB WITH SAME DATA (Y/N) ?
```

The printout contains, columnwise, the number of successes, the theoretical probabilities p_i, the expected frequencies $e_i = p_i N$, the observed frequencies o_i, and the standardized residuals $z_i = (e_i - o_i)/e_i^{1/2}$. In our example we obtain noteworthy discrepancies for four and five successes. The overall Pearson-X^2 has a tail probability of $p < 0.01$ ($df = 1$). Thus, we may conclude that the assumption that the patient sample can be decomposed into two samples from two populations is untenable; therefore, we calculated an alternative model in which we assumed three rather than two patient populations. The parameter estimates for this model appear below.

```
PROGRAM BINOMIX       DATE = 10-25-1989       TIME = 15:48:17       PAGE = 16
JOB TITLE : Workbook Data Example Finite Binomial Mixtures
-----------------------------------------------------------------------------

RESULTS OF PARAMETER ESTIMATION FOR UNIVARIATE LATENT-CLASS-MODEL
WITH LOCAL BINOMIAL DISTRIBUTIONS:

    CLASS          P(CLASS)     P(SUCCESS / CLASS)
-----------------------------------------------------
      1            0.59192          0.55530
      2            0.04940          0.99961
      3            0.35868          0.07507
-----------------------------------------------------

LAST MAXIMUM DEVIATION IN PARAMETER ESTIMATION : 9.80329974453742D-06
NUMBER OF ITERATIONS : 464

            ( PRESS ANY KEY TO CONTINUE )
```

The parameter estimates suggest that the three classes differ considerably in size. The third class contains 35.77% of the population, the first 59.16%, and the second only 5%. The success rate in the third class is only 7.51%; in the first class it is 55.53%. In the second class success is almost certain. Looking at the conditional probabilities in the next printout, we realize that patients freed of no symptom or only one most likely belong to class 3. Patients freed of three, four, or five symptoms most likely belong to class 1. Patients freed of all symptoms belong to the smallest, the second class.

```
PROGRAM BINOMIX        DATE = 10-25-1989        TIME = 15:49:40        PAGE = 17
JOB TITLE : Workbook Data Example Finite Binomial Mixtures
--------------------------------------------------------------------------------

CONDITIONAL PROBABILITIES P(CLASS / SUCCESSES) :

SUCCESSES I  CLASS 1   CLASS 2   CLASS 3
----------------------------------------
    0      I  0.04067   0.00000   0.95933
    1      I  0.39479   0.00000   0.60521
    2      I  0.90939   0.00000   0.09061
    3      I  0.99357   0.00000   0.00643
    4      I  0.99879   0.00079   0.00042
    5      I  0.38797   0.61202   0.00001
----------------------------------------

              ( PRESS ANY KEY TO CONTINUE )
```

The evaluation of the model shows that the addition of the third class led to a much closer approximation of the observed by the expected frequencies, in particular for four and five successes. Accordingly, the goodness-of-fit *chi*-square statistics are very small and indicate an excellent fit. The evaluation summary appears below.

```
PROGRAM BINOMIX        DATE = 10-25-1989        TIME = 15:50:13        PAGE = 18
JOB TITLE : Workbook Data Example Finite Binomial Mixtures
--------------------------------------------------------------------------------

EVALUATION OF LATENT-CLASS-MODEL WITH 3 LATENT CLASSES:

SUCCESSES  THEOR. PROB.   EXPECTED N   OBSERVED N   STD. RESID.
--------------------------------------------------------------
    0        0.25310        91.12        91.00        -0.01
    1        0.16280        58.61        59.00         0.05
    2        0.17651        63.54        62.00        -0.19
    3        0.20174        72.63        75.00         0.28
    4        0.12530        45.11        44.00        -0.16
    5        0.08056        29.00        29.00        -0.00
--------------------------------------------------------------

FIT INDICES:
PEARSON CHI-SQUARE            =     0.1450
LIKELIHOOD-RATIO CHI-SQUARE =     0.1446

NEW JOB WITH SAME DATA (Y/N) ?
```

The program BINOMIX is very convenient to use. It is fully interactive and can be controlled from the keyboard; however, there are a few constraints. First, in the program's present form the keyboard is the only way to input data. Second, there is no option to send the output to the printer; however, the results appear conveniently organized in units of one screen each. Thus, they can be sent to the printer using the "print screen" key, and no overlap in the printout will result. Third, the program does not give the degrees of freedom for the chi-square values, nor does it give tail probabilities.

One problem is that one has to make the assumption that the probabilities of the m categories are equal. If this is not the case, programs for latent-class analysis can be used.

There are only a few programs that perform analysis of finite-mixture models. The program BINOMIX can be obtained from Dr. E. Erdfelder, Department of Psychology, University of Bonn, Römerstr. 164, D-5300 Bonn 1, West Germany.

If one is willing to confine oneself to the analysis of mixtures of multinomial distributions, programs for latent-class analysis can be used (see chapter on LCA).

Chapter 13

Prediction Analysis of Cross-Classifications

1. Summary of Method

Prediction analysis of cross-classifications is a method for analyzing point predictions in two- or higher-dimensional contingency tables (*cf.* Hildebrand *et al.*, 1977; Szabat, 1990; von Eye and Brandtstädter, 1988a). The method requires the researcher to classify variables in predictors and criteria. Typically, the analysis proceeds in the following four steps:

1. Formulation of prediction hypotheses. To formulate the hypotheses, prediction logic (Hildebrand *et al.*, 1977; Szabat, 1990) and statement calculus (von Eye and Brandtstädter, 1988a) have been used. These formalized approaches allow one to avoid redundancies, to identify contradictory hypotheses, and to identify the cells containing cases meeting with the predictions (hit cells) and the cells containing cases contradicting the predictions (error cells).

2. Estimating expected frequencies against which the hypotheses can be tested. Cast in log-linear modeling terminology, the hierarchical model used for estimating the expected frequencies is [predictors], [criteria]. In other words, the expected frequencies are estimated to meet with the following assumptions:

201

—there may be interactions of any order between the predictors,

—there may be interactions of any order between the criteria,

—predictors and criteria are independent of each other.

These assumptions can be violated only if there are relationships between predictors and criteria. The prediction hypotheses tested by prediction analysis are examples of such relationships.

3. Calculation of descriptive statistics. The descriptive statistics best known in prediction analysis include a measure for proportionate reduction in error (PRE), the scope, and the precision. PRE, or Del tells us how many more cases contradict the prediction hypothesis in the expected than in the observed frequency distribution. The scope of a prediction is the proportion of the population included in non-tautological predictions. A prediction's precision is defined as the number of errors observed. In tautological predictions which cannot be contradicted, the number of prediction errors is zero. Later, we give a numerical example of a tautological prediction.

Other descriptive measures for the prediction success include measures for the proportional increase in hits (PIH; see von Eye and Brandtstädter, 1988a). These measures are constructed parallel to PRE; however, they focus on hit cells. A large PIH indicates that the prediction hypothesis identifies more hits in the observed frequency distribution than was estimated from the expected frequency distribution. In particular, the following two PIH measures have been discussed (see von Eye and Brandtstädter, 1988a):

$$PIH1 = (S1 - S2)/S2 \qquad (1)$$

and

$$PIH2 = (S1 - S2)/(n - S2) \qquad (2)$$

where $S1$ denotes the sum of the frequencies in the hit cells in the observed frequency distribution and $S2$ denotes the sum of the frequencies in the hit cells in the expected frequency distribution. Multiplied by 100, PIH1 gives the percentage of increase in hits when the observed frequencies in the hit cells are compared with their expected frequencies. PIH2 expresses this increase relative to the maximal increase. If there are no tautological predictions, PIH2 is identical to PRE, otherwise, PRE and PIH2 differ.

4. Inferential evaluation of prediction success. To statistically evaluate PRE and PIH measures z-tests (Hildebrand et al., 1977; Szabat, 1990) and binomial tests (von Eye and Brandtstädter, 1988a) have been discussed.

More recent developments of prediction analysis include methods for analysis of ordinal variables (von Eye and Brandtstädter, 1988b), methods for analysis of ordinal hypotheses (Rudinger et al., 1985), and allow one to evaluate the bias in exploratory prediction analysis (Szabat, 1982). The following sections discuss computational procedures for the steps 2 to 4 of prediction analysis.

2. Outline of the Computational Procedures

This section outlines the computational procedures for computation of a prediction analysis using a microcomputer. In the next section, these steps will be illustrated in detail using a data example. For preparation of data SYSTAT (Wilkinson, 1988) will be used. A program by von Eye and Krampen (1987) will be used to perform the prediction analysis.

For the data preparation step we assume the raw data are available in a file. To do the preprocessing, we determine what variables are predictors and what variables are criteria. Then we cross-tabulate predictors with criteria to obtain the two-dimensional contingency table expected as input for the program used below. If there is more than one predictor, a composite predictor variable must be formed that contains each configuration of the original predictors as a state. Consider the two predictors A and B. A has levels 1, 2, and 3. B has levels 1 and 2. The cross-tabulation of A and B has the following cells: 11, 12, 21, 22, 31, and 32. The composite predictor also has levels 11 to 32, and the predictions are made from these configurations.

The criteria must be processed in the same way. If the expected frequencies are estimated using programs for log-linear modeling and the statistical analysis is handcalculated, these transformations are not necessary. Rather, expected frequencies are estimated under a model that assumes two groups of variables. The first group is formed by the predictors, the second by the criteria. As was specified above, the model assumes that, within each group of variables, any type of interaction may prevail; however, it assumes independence between the groups.

Data can be fed into the program either via files or interactively. The program expects or prompts the following input information: (1) size of the matrix, (2) observed cell frequencies, and (3) hypothesized hit and error cell pattern. For models including ordinal variables the ranks of the variable states can be considered also.

The program was compiled and tested on an IBM compatible HP100 Plus. It should operate under DOS 2.1 or higher. After data entry, the program sends the results directly to the printer. Other hit and error cell patterns or other data sets can be analyzed without restarting the program.

3. An Annotated Data Example

The following data example involves one predictor and one criterion. Consider the variables Ordination (*O*) and Number Conservation (*C*) in a study on childhood development (*cf.* Froman and Hubert, 1980). Each child's raw data contain a measure for the observed state in *O* and a measure for the observed state in *C*. For the following steps we assume these data are available on the ASCII file OC.DAT.

Before performing prediction analysis, one must cross-tabulate *O* and *C*. The statistical software package SYSTAT (Wilkinson, 1988) requires the following steps to transform the raw data into a cross-tabulation:

Step 1: Preparation of Data

1. Generating a SYSTAT system file. To be able to apply statistical programs from the SYSTAT package, raw data must be transformed into a system file. Raw data files must be written in ASCII code and must have the suffix .DAT. To generate the system files one uses the DATA module. Within this module one performs the following operations:

SAVE OC (Initiates the saving of a system file named OC.SYS)

INPUT O,C (Defines the variables to be used)

GET OC (Reads the ASCII file with the raw data)

RUN (Saves the data as a SYSTAT system file).

2. Generating the cross-tabulation for the prediction analysis. The TABLES module in SYSTAT allows one to transform raw data into cross-tabulations. Within this module one performs the following operations:

USE OC (Reads the raw data from system file)

OUTPUT@ (Sends output to printer; otherwise, it will appear only on the screen)

TABULATE O*C (Generates cross-tabulation; result is sent to both printer and screen).

SYSTAT allows one to save cross-tabulations; however, the program used for prediction analysis cannot read these tables.

The result of the cross-tabulation reads as follows:

```
TABLE OF        A      (ROWS) BY        B       (COLUMNS)

FREQUENCIES

            1.000     2.000     3.000      TOTAL
         Öáááááááááááááááááááááááááááááá¢
  1.000 ·      16         3         1   ·      20
         ·                                ·
  2.000 ·      15         3         3   ·      21
         ·                                ·
  3.000 ·      23         4        27   ·      54
         ãáááááááááááááááááááááááááááááì
  TOTAL        54        10        31          95
```

Step 2: Data Input

The program by von Eye and Krampen (1987) allows one to read cross-tabulations from ASCII files or to key them in interactivley. The files must contain the numbers of rows and columns and the raw frequencies of the Predictors x Criteria cross-tabulation. These numbers must be separated by blanks or commas. Files for the hit and error cell patterns must contain only the zeros for the error and the ones for the hit cells. Examples of matrices containing the observed frequencies and a matrix containing a hit-cell pattern follow below (adapted from Froman and Hubert, 1980).

```
3 3
16 3 1
15 3 3
23 4 27

1 0 0
1 1 0
1 1 1
```

If the user decides to key in the frequencies or the hit and error cell pattern, the program asks for this information cell-wise. After completion of the data input the program provides an option for correcting the data from the keyboard.

Step 3: Running the Program

The program will send the results to the printer. A sample printout is given below.

```
Prediction Analysis
BASIC program; author: Alexander von Eye
Ordination and Number Conservation

observed frequencies
-------- -----------

   16    3    1    20
   15    3    3    21
   23    4   27    54

   54   10   31    95

N = 95
```

```
expected frequencies
-------- -----------

11.37   2.11   6.53
11.94   2.21   6.85
30.69   5.68  17.62


matrix of hit cells
------ -- --- -----

1 0 0
1 1 0
1 1 1
```

```
there is weak evidence in support of the hypothesis
PIH1 =  .3325083    PIH2 =  .1221027    z =  1.963928
precision = .1629917  scope =  .431579
p(z) = 0.024769
DEL =  .5479266
```

```
chisquare for model [predictors][criteria]: chi2 =  17.60075
g-square for same model: g2 =  19.99556
df = 4
```

```
evaluation of partial hypotheses
---------- -- ------- ----------

partial hyp. fo    fe    del   precis.  del(cum)
------------ --    --    ---   -------  --------

           1   4   8.63  0.54   0.09     0.30
           2   3   6.85  0.56   0.07     0.55
           3   0   0.00  0.00   0.00     0.55
```

An optional title and name for the run may be keyed in by the user. The program prints the matrix of the observed frequencies and, if corrections were made, the matrix of the corrected observed frequencies. These are followed by the expected frequencies, calculated under the log-linear model [predictors], [criteria]. This matrix is followed by the hit and error cell pattern. A 1 denotes a hit cell, a 0 denotes an error cell.

The results are printed in two sections. The first contains the descriptive measures and the test statistics. For this example we realize that there is weak evidence in support of the hypothesis. (For an explanation of the terms "strong," "weak," and "neither strong nor weak support" see von Eye and Brandtstädter, 1988a.) Then we see that the hit-cell pattern accounts for 33.25% more hits than the assumption of independence between predictors and criteria (PIH1), that is, 12.21% of the maximally possible increase in hits (PIH2). The z-statistic for this result, estimated under a binomial model, is 1.96. Both the precision and the scope are relatively low. There is no test statistic comparing these values with some null hypothesis. The tail probability $p(z)$ indicates that the increase in hits is significant. The reduction in error measure Del indicates that the error cell pattern of the hypothesis allows the researcher to make 54.79% fewer errors than the assumption of independence of predictors and criteria.

In this example PIH2 and Del are different. The reason is that for the third level of the predictor a tautological prediction is made. For this predictor level any outcome may occur; therefore, one cannot discriminate among the three criteria levels, and the prediction cannot be empirically nor logically contradicted. The program, therefore, eliminates such cells from the set of hit cells and calculates prediction success from the remaining hit cells. The error cell pattern is not affected by this type of tautological predictions.

The second section of the results contains a decomposition of the overall Del. For each row of the cross-tabulation a partial Del and the precision are calculated using the observed and expected frequencies in the error cells. For each row the contribution to the overall Del is given also. The example shows that for the tautological prediction in the last row of the matrix no error cells occur, and thus no reduction in prediction errors occurs.

In addition to a standard confirmatory prediction analysis, the program allows one to test prediction hypotheses for ordinal variables, that is, variables with ranks as categories. To illustrate this procedure, we recalculated our example under the assumption that both the predictor's and the criterion's categories have the ranks 1, 2, and 3. The following printout shows the results of the analysis.

```
transformed ranks of predictor
 .3333334  .6666667  1
transformed ranks of criterion
 .3333334  .6666667  1

expected frequencies for ordinal model
-------- ----------- --- ------- -----

   15.87    3.13    0.99   20.00
   15.26    2.57    3.17   21.00
   22.86    4.29   26.85   54.00
   54.00   10.00   31.01    0.00
# of iterations IT =  5


-----------------------------------------------------
there is weak evidence in support of the hypothesis
PIH1 =  8.719444E-03    PIH2 =  4.794923E-03    z =  6.302281E-02
precision = 7.677467E-02  scope =  .431579
p(z) = 0.474874
DEL =  4.025364E-02
-----------------------------------------------------
chisquare for model [predictors][criteria]: chi2 =  .1127964
g-square for same model: g2 =  .1104705
df =  1
-----------------------------------------------------
evaluation of partial hypotheses
---------- -- ------- ----------

partial hyp.  fo    fe    del  precis.  del(cum)
------------ --    --    ---  -------  --------

          1    4  4.13   0.03   0.04    0.02
          2    3  3.17   0.05   0.03    0.04
          3    0  0.00   0.00   0.00    0.04
-----------------------------------------------------
```

The printout gives the degrees of freedom for the log-linear model that considers the main effects, and the ranks of the variables are given. The expected frequencies for ordinal models are typically closer to the observed frequencies than in standard prediction analysis. In the present example they are so close that the overall goodness-of-fit indices are only 0.11.

If the differences between observed and expected frequencies are small, the prediction success cannot be colossal. The printout shows that only 0.0087% hits are gained, that is less than one individual and only 0.0047% of the maximally possible gain. Accordingly, the other descriptive measures and the statistical test indicate that the prediction success under the ordinal model is less than impressive, thus reflecting the general result that an increase in the amount of information used for the estimation of expected frequencies typically leads to a reduction in the differences between expected and observed frequencies. Only the scope remains unaffected.

There are only a few programs available that perform prediction analysis. The program used to calculate the examples in this chapter has been published by von Eye and Krampen (1987). It can be obtained from Alexander von Eye, The Pennsylvania State University, College of Health and Human Development, S-110 Henderson South, University Park, PA 16802. (Please send in a diskette.)

Other options to calculate prediction analysis include estimating expected frequencies using a program for log-linear modeling (BMDP, SPSSX, SAS, SYSTAT). The model for estimation of expected frequencies is [predictors], [criteria]. The estimates from such a model can be used for hand-calculation of all descriptive parameters and the z-test.

Chapter 14

Configural Frequency Analysis

1. Summary of Method

Configural Frequency Analysis (CFA; Krauth and Lienert, 1973; von Eye, 1990a, 1990b) is a method for typological search in cross-classifications. CFA inspects each cell in a contingency table and tests whether

$$o \neq e$$

that is whether the observed cell frequency is different from the estimated expected cell frequency. Individuals in a cell for which

$$o > e$$

belong to a type. Individuals in a cell for which

$$o < e$$

belong to an antitype. Both types and antitypes are characterized by the pattern of categories that define the cell. Types indicate that these categories go well together. Antitypes indicate that the categories do not go well together.

The estimation of expected frequencies uses methods from log-linear modeling. Models used for CFA include the uniform-distribution model (zero-order CFA), the main-effect model (first-order CFA), and the model of independence between two sets of variables which, within each set, may interact (Interaction Structure Analysis, Prediction CFA). In all models of CFA, types or antitypes suggest that, locally, the model used for estimating expected frequencies does not hold true. Specifically, the presence of types or antitypes suggests

1. In *zero-order CFA*: there are at least main effects that is the assumption of random assignment of variable categories is untenable. Types and antitypes in zero order CFA can be interpreted as clusters and as sparse sectors in the data space, respectively.

2. In *first order CFA*: there are local associations between variables. Examples of types in first order CFA include syndromes that is characteristic combinations of symptoms as "the midlife crisis." Examples of antitypes include combinations of symptoms that do not go well together such as, for instance, "hilarious depression."

3. In *Interaction Structure Analysis*: there are local associations between independent and dependent variables. Types suggest that a particular combination of states of the independent variables allows one to expect a particular combination of states of the dependent variables. Antitypes suggest that certain combinations of dependent variable states are unlikely to follow certain combinations of independent variable states.

In exploratory CFA, testing proceeds in three steps. First, one calculates the expected frequencies e_i and the test statistics, for instance, chi-square values. Second, one adjusts the significance level alpha to the number of tests performed. This number typically equals the number of cells in the contingency table, t. Well-known strategies for *alpha* adjustment include Bonferroni's and Holm's (1979) methods. Bonferroni's method yields the adjusted alpha

$$\text{alpha}^* = \text{alpha}/t. \qquad (1)$$

The more efficient Holm method yields for the ith test

$$\text{alpha}_i^* = \text{alpha}/(t - i), \text{ for } i = 0, ..., t\text{-}1 \qquad (2)$$

where alpha is the *a priori* determined alpha and i indexes the number of significance tests already performed before each test. We obtain, for example, $i = 0$ for the first test, $i = 1$ for the second test, and $i = t - 1$ for the last test. For Holm's method, the test statistics must be rank-ordered such that the cell with the smallest tail probability (see below) is tested first, and so on. Testing terminates when, for the first time, a null hypothesis cannot be rejected. Even more efficient methods have recently been discussed by Holland and Copenhaver (1987).

In the third step of exploratory CFA, most researchers use Pearson chi-square components, defined as

$$X^2 = (o - e)^2 / e, \tag{3}$$

or the binomial test. More powerful tests have been discussed by Lehmacher (1981; *cf.* Küchenhoff, 1986; von Eye and Rovine, 1988).

2. Outline of the Computational Steps

This section outlines the steps for computing CFA using a PC. In the next section, these steps will be illustrated in detail using a data example. A program by von Eye (1990) will be used to perform the calculations.

When applying this program, data must be categorical. If variables must be categorized, application of a program for data categorization is necessary. The CFA program expects cell frequencies as input. Therefore, the variables must be cross-tabulated. Cross-tabulation can be done with any of the major statistical software packages for PCs (*e.g.*, SYSTAT, SAS, SPSSX, BMDP).

The CFA program can process data read from a file or data keyed in. The program expects the following information: (1) size of cross-tabulation, (2) (only when data are read from file) cell indices, and (3) observed cell frequencies. Data on files are expected to be separated by commas or blanks.

After data entry, the program prompts the user to make a decision as to what significance test to use. The program allows one to analyze another data set without need to restart.

3. An Annotated Data Example

The following example analyzes responses to two parallel forms of a state anxiety questionnaire, each administered four times (Nesselroade *et al.*, 1986; von Eye and Nesselroade, 1990). For the following steps we assume each subjects' scale values are available for each observation point.

Before application of CFA, these scale values must be categorized in a sensible manner. There are several ways to transform and categorize longitudinal data. Examples include the calculation of first or second differences, that is, differences between the original, time-adjacent data points and differences between differences. One well-known effect of calculating first differences is the elimination of the linear trend (See chapter on Spectral Analysis.)

For the present example, we approximated the short-time series using orthogonal polynomials. Orthogonal polynomials can be described as systems of polynomials that have the following form:

$$y = a_o X_o^0 + a_1 X_1^1 + a_2 X_2^2 + \dots \tag{1}$$

where X_i^i is a polynomial of the ith order, for $i = 0, 1, 2, \dots$ For instance, a first-order polynomial is given by

$$y = a_o + a_1 x_1, \tag{2}$$

that is a standard regression formula.

The polynomial coefficients can be estimated and interpreted as ordinary least-squares regression coefficients: a_o is the arithmetic mean of the time series, a_1 indicates the linear trend, and a_2 the curvature. These coefficients can be categorized and analyzed with CFA. (Alternatively, they can be analyzed with ANOVA.) A natural cutoff point for polynomial coefficient a_o is the median. A natural cutoff for coefficient a_1 is zero, with coefficients $a_1 > 0$ indicating a positive slope and coefficients $a_1 < 0$ indicating a negative slope. The coefficients for the quadratic trend can also be categorized at zero, with positive coefficients indicating a U-shaped curvature and negative coefficients indicating an inversly U-shaped curvature.

Most statistical software packages allow one to calculate coefficients for orthogonal polynomials. In the following example we assume these coefficients have been calculated (For an example of the use of orthogonal polymonials see the chapter on growth curve analysis.), and for each subject the pattern of categorized coefficients is stored on file RAW.DAT. For the following preparatory steps we use SYSTAT (Wilkinson, 1988).

Step 1: Preparation of Data

1. *Generating a SYSTAT system file for the raw data.* SYSTAT programs can process data only if they are read from a system file. For the following transformations we assume the raw data (categorized polynomial coefficients) are stored in the ASCII file RAW.DAT. (For other alternatives for data entry with SYSTAT see the chapter on Prediction Analysis.) For each subject we have the measures $A11, A12, A21,$ and $A22$, where $A11$ is the categorized polynomial coefficient a_o for the first anxiety scale, $A12$ is the categorized linear trend, a_1, for the first scale, $A21$ is the categorized polynomial coefficient a_o for the second anxiety scale, and $A22$ is the categorized linear

trend a_1 for the second scale. The commands that transform the ASCII raw data file into a SYSTAT system file are as follows:

SAVE RAW (Initiates saving of data in system file RAW.SYS.)

INPUT A11,A12,A21,A22 (Names the variables to be used.)

GET RAW (Reads the raw data from ASCII file RAW.DAT.)

RUN (Starts the saving of the data.) The command SWITCHTO TABLES transfers us to the TABLES module.

2. *Generating the cross-tabulation for CFA.* The TABLES module in SYSTAT allows one to form multivariate contingency tables from raw data. Within this module one applies the following commands:

USE RAW (Reads raw data from system file RAW.SYS.)

OUTPUT @ (Sends output to printer.)

TABULATE A11*A12*A21*A22 (Generates four-dimensional cross-tabulation; output is sent to both printer and screen.)

Using SYSTAT one can save cross-tabulations on files. However, the program used for CFA cannot read these tables. For the present example, the cross-tabulation generated by SYSTAT is printed as follows:

```
TABLE OF        A21      (ROWS) BY      A22      (COLUMNS)
     FOR THE FOLLOWING VALUES:
                   A12      =    1.000
                   A11      =    1.000

FREQUENCIES

                 1.000      2.000      TOTAL
           Öáááááááááááááááááááá¢
   1.000 ·      12         0  ·        12
         ·                    ·
   2.000 ·       3         0  ·         3
           ááááááááááááááááááááì
TOTAL          15         0            15

TABLE OF        A21      (ROWS) BY      A22      (COLUMNS)
     FOR THE FOLLOWING VALUES:
                   A12      =    2.000
                   A11      =    1.000

FREQUENCIES

                 1.000      2.000      TOTAL
           Öáááááááááááááááááááá¢
   1.000 ·       0         6  ·         6
         ·                    ·
   2.000 ·       1         4  ·         5
           ááááááááááááááááááááì
TOTAL           1        10            11
```

```
TABLE OF        A21     (ROWS) BY        A22     (COLUMNS)
     FOR THE FOLLOWING VALUES:
                 A12      =     1.000
                 A11      =     2.000

FREQUENCIES

                1.000    2.000      TOTAL
              Öáááááááááááááááááááá¢
    1.000  .     0        0    .      0
           .                   .
    2.000  .     6        0    .      6
              ââââââââââââââââââââì
    TOTAL        6        0           6


TABLE OF        A21     (ROWS) BY        A22     (COLUMNS)
     FOR THE FOLLOWING VALUES:
                 A12      =     2.000
                 A11      =     2.000

FREQUENCIES

                1.000    2.000      TOTAL
              Öáááááááááááááááááááá¢
    1.000  .     0        1    .      1
           .                   .
    2.000  .     1        8    .      9
              ââââââââââââââââââââì
    TOTAL        1        9          10
```

Step 2: Data Input

The CFA program allows one to read cross-tabulations from ASCII files or to key in data interactively. Files must contain the following information:

1. Number of variables in the cross-tabulation (not counting the frequencies as a variable),
2. for each variable the number of categories, and
3. the cell indices and cell frequencies.

All values must be separated by blanks or commas. The following printout shows the cross-tabulation read in for the state anxiety example. The file was generated as an ASCII file using a word processor. The program prompts for the name of the data file.

```
4  2  2  2  2
        1.000,      1.000,      1.000,      1.000,      12.000
        1.000,      1.000,      1.000,      2.000,       0.000
        1.000,      1.000,      2.000,      1.000,       3.000
        1.000,      1.000,      2.000,      2.000,       0.000
        1.000,      2.000,      1.000,      1.000,       0.000
        1.000,      2.000,      1.000,      2.000,       6.000
        1.000,      2.000,      2.000,      1.000,       1.000
        1.000,      2.000,      2.000,      2.000,       4.000
        2.000,      1.000,      1.000,      1.000,       0.000
        2.000,      1.000,      1.000,      2.000,       0.000
        2.000,      1.000,      2.000,      1.000,       6.000
        2.000,      1.000,      2.000,      2.000,       0.000
        2.000,      2.000,      1.000,      1.000,       0.000
        2.000,      2.000,      1.000,      2.000,       1.000
        2.000,      2.000,      2.000,      1.000,       1.000
        2.000,      2.000,      2.000,      2.000,       8.000
```

The first row of numbers contains the information on the cross-tabulation under study. We are analyzing four variables with two levels each. The cross-tabulation in tabular form follows. If the user chooses to key in the data, the program prompts this information. After completion of the data input, the program provides an option for data correction from the keyboard.

Step 3:   Running the Program

The program sends results to the printer. A sample printout is given below.

```
BASIC-program CFA
author of program: Alexander von Eye, 1988
limitations: 1 < nvar < 10
             # states per variable < 100
input: either interactive or via file


sample size n= 42
alpha*= .003125


Anscombe's test will be performed




marginals of variable 1
--------- -- -----------

 26   16

marginals of variable 2
--------- -- -----------

 21   21

marginals of variable 3
--------- -- -----------

 19   23

marginals of variable 4
--------- -- -----------

 23   19


table of results
----- -- -------


configuration    fo   fe        test statistic p(z)
-----------------------------------------------------------------
 1  1  1  1      12   3.220522   3.871625      5.407678329649727D-05
 1  1  1  2       0   2.660431   2.343345      9.555837785980716D-03
 1  1  2  1       3   3.898526   .3896048      .3484143958505687
 1  1  2  2       0   3.220522   2.598179      4.686010585779182D-03
 1  2  1  1       0   3.220522   2.598179      4.686010585779182D-03
 1  2  1  2       6   2.660431   1.864245      3.114357464146129D-02
```

```
1  2  2  1       1   3.898526  1.681018   4.637966111679575D-02
1  2  2  2       4   3.220522   .5121774   .3042634278653689
2  1  1  1       0   1.98186   1.991564   2.320938718887651D-02
2  1  1  2       0   1.637188  1.786717   3.699157805041917D-02
2  1  2  1       6   2.399093  2.066266   1.940162581918625D-02
2  1  2  2       0   1.98186   1.991564   2.320938718887651D-02
2  2  1  1       0   1.98186   1.991564   2.320938718887651D-02
2  2  1  2       1   1.637188   .4050348   .3427259503549101
2  2  2  1       1   2.399093   .9180286   .179301896755918
2  2  2  2       8   1.98186   3.361952   3.870218917786189D-04
```

The first part of the printout gives general information on the program. This part is followed by information on the number of variables, the sample size, and the Bonferroni-adjusted alpha, based on alpha = 0.05. For the present example, we obtain alpha = 0.05/16 = 0.003125.

Next, the program specifies the significance test applied for type and antitype testing. The user has the choice between the following eight tests:

1. Binomial test,
2. Binomial test approximated using Stirling's formula,
3. Pearson's chi-square component test,
4. z-test, approximated through (1),
5. z-test, approximated through (3),
6. Lehmacher's approximative hypergeometric test,
7. Lehmacher's approximative hypergeometric test with Küchenhoff's continuity correction, and
8. Anscombe's (1953) chi-square test.

These tests differ in power and applicability when sample sizes are small. For instance, Lehmacher's test is the most powerful; it requires, however, large samples. Anscombe's test is said to approximate the *chi*-square distribution better than the other chi-square tests, and Pearson's test performs better than other approximative tests when sample sizes are small. (For more details see von Eye, 1990; von Eye and Rovine, 1988.)

The next part of the printout contains the univariate marginal sums. The last part contains the results of the cell-wise CFA tests.

For the present example, we used Anscombe's test. This test uncovered two types and no antitype. The first type has pattern 1111. It suggests that more subjects than expected by chance show above average state anxiety and a linear increase in state anxiety in both test forms. The second type has pattern 2222, suggesting that more subjects than expected by chance show below average state anxiety and a linear decrease in state anxiety, also in both test forms. Together, these two types suggest that the two test forms are parallel.

There are only a few programs available that do CFA. The program used for the present examples can be obtained as a BASIC source code tested using DOS PCs from Alexander von Eye, The Pennsylvania State University, College of Health and Human Development, S-110 Henderson S., University Park, PA 16802.

Please send in a diskette, and you will receive the program gratis. If you use one of the major statistical software packages, you can simulate a CFA run by estimating a main-effect log-linear model and requesting the standardized (or some other form of the) residuals. Standardized residuals can be considered z-scores or, if squared, chi-square values with one degree of freedom.

References

AFFIFI, A., AND CLARK, V. (1984). *Computer-aided multivariate analysis.* Belmont, CA: Lifetime Learning Publications.

ALLISON, P. D. (1984). *Event history analysis: Regression for longitudinal event data.* Beverly Hills: Sage.

ANDERSON, T. W. (1958). *An introduction to multivariate statistical analysis.* New York: Wiley.

ANSCOMBE, F. J. (1953). Contribution to discussion of paper by H. Hotelling "New light on the correlation coefficient and its transform." *Journal of the Royal Statistical Society. B.* **15**, 229–230.

ARABIE, P., CARROLL, J. D., AND DESARBO, W. S. (1987). *Three-way scaling and clustering.* Newbury Park CA: Sage.

ARBUCKLE, J. (1989). *AMOS users manual.* Philadelphia: Temple University.

BAKER, R. J., AND NELDER, J. A. (1978). *The GLIM system, Release 3, Generalized linear interactive modeling.* Numerical Algorithms Group, Oxford.

BELSKY, J., AND ROVINE, M. (1990). "Patterns of marital change across the transition to parenthood," *Journal of Marriage and the Family.* **52**, 5–19.

BELSKY, J., GILSTRAP, B., AND ROVINE, M. (1984). "The Pennsylvania Infant and Family Development Project. I: Stability and change in mother–infant and father–infant interaction in a family setting at 1-, 3-, and 9-months," *Child Development*, **55**, 692–705.

BENTLER, P. (1985). *EQS user's manual.* Los Angeles: BMDP Statistical Software.

BENTLER, P., AND BONNET D. (1980). "Significance tests and goodness of fit in the analysis of covariance structures," *Psychological Bulletin.* **88**(3), 588–606.

BISHOP, Y. M. M., FIENBERG, S. E., AND HOLLAND, P. W. (1975). *Discrete multivariate analysis. Theory and practice.* Cambridge, MA: The MIT Press.

BLOSSFELD, H. P., HAMERLE, A., MAYER, K. U. (1988). *Event history analysis.* Hillsdale, NJ: Erlbaum.

BMDP STATISTICAL SOFTWARE (1985). Berkeley, CA: University of California Press.

BOCK, R. D. (1975). *Multivariate statistical models in behavioral research.* New York: McGraw-Hill.

BOLLEN, K. (1989). *Structural equations modeling with latent variables.* New York: Wiley.

BORG I (Ed.). (1981). *Multidimensional data representations: When and why?* Ann Arbor, MI: Mathesis Press.

BOX, G. E. P., AND JENKINS, G. M. (1976). *Time series analysis: Forecasting and control.* San Francisco: Holden Day, Inc.

BROCKWELL, P. J., AND DAVIS, R. A. (1987). *Time series: Theory and methods.* New York: Springer.

BROCKWELL, P. J., DAVIS, R. A., AND MANDATINO, J. V. (1987). *Software to accompany the book by Brockwell and Davis.* New York: Springer.

BRILLINGER, D. R. (1975). *Time series analysis: Data analysis and theory.* New York: Holt, Rinehart, and Winston.

BURCHINAL, M., AND APPELBAUM, M. (1987). *Estimating individual developmental functions: A discussion of various methods and their assumptions.* Chapel Hill, NC: L. L. Thurstone Laboratory, University of North Carolina. (No. 178).

CARROLL, J. D., AND CHANG, J. J. (1970). "Analysis of individual differences in multidimensional scaling via an N-way generalization of the 'Eckard–Young' decomposition," *Psychmetrika.* **35**, 283–319.

CLOGG, C. C. (1977). *Unrestricted and restricted maximum likelihood latent structure analysis: A manual for users.* The Pennsylvania State University: Population Issues Research Office, unpublished manual.

CLOGG, C. C. (1981). New developments in latent structure analysis. In D. M. Jackson and E. F. Borgatta (Eds.), *Factor analysis and measurement in sociological research* (pp. 214–280). Beverly Hills, CA: Sage.

CLOGG, C. C. (1988). "Latent class models for measuring." In R. Langeheine and J. Rost (Eds.), *Latent trait and latent class models* (pp. 173–205). New York: Plenum.

CLOGG, C. C., AND ELIASON, S. R. (1987). "Some common problems in log-linear analysis," *Sociological Methods and Research.* **16**, 8–44.

CLOGG, C. C., ELIASON, S. R., AND GREGO, J. (1990). Models for the analysis of change in discrete variables. In A. von Eye (Ed.), *Statistical methods in longitudinal research* (Vol. 2). Boston: Academic Press.

COOK, T. D., AND CAMPBELL D. T. (1979). *Quasi-experimentation: design and analysis issues for field settings.* Chicago: Rand McNally.

COX, D. R. (1972). "Regression models and life tables," *Journal of Royal Statistical Society. B.* **34**, 187–220.

DAVIS, R. B. (1983). Optional inference in the frequency domain. In D. R. Brillinger and P. R. Krishnaiah (Eds.), *Time series in the frequency domain* (pp. 73–92). Amsterdam: North Holland.

DEMPSTER, A. P. (1969). *Elements of continuous multivariate analysis.* Reading, MA: Addison-Wesley.

DEMPSTER, A. P., LAIRD, N. M., AND RUBIN, D. B. (1977). "Maximum likelihood estimation from incomplete data via the EM-algorithm," *Journal of the Royal Statistical Society. B.* **39**, 1–22.

ELIASON, S. R. (1988). *The categorical data analysis system (CDAS).* The Pennsylvania State University: Department of Sociology, unpublished manual.

ERDFELDER, E. (1990). Deterministic developmental hypotheses, probabilistic rules of manifestation, and the analysis of finite mixture distributions. In A. von Eye (Ed.), *Statistical methods in longitudinal research* (Vol. 2). Boston: Academic Press.

EVERITT, B. S., AND HAND, D. (1981). *Finite mixture distributions.* London: Chapman and Hall.

EVERITT, B. S., AND DUNN, G. (1988). Log-linear modeling, latent class analysis, or correspondence analysis: Which model should be used for the analysis of categorical data? In R. Langeheine and J. Rost (Eds.), *Latent trait and latent class models* (pp. 109–127). New York: Plenum.

FIENBERG, S. E. (1980). *The analysis of cross-classified categorical data.* Cambridge, MA: The MIT Press.

FRASER, C. AND MCDONALD, R. P. (1988). "COSAN: Covariance Structure Analysis," *Multivariate Behavioral Research.* **23**, 263–265.

FROMAN, T., AND HUBERT, L. J. (1980). "Application of prediction analysis to developmental priority," *Psychological Bulletin.* **87**, 136–146.

GAMES, P. (1990). Alternative analyses of repeated measures designs by ANOVA and MANOVA. In A. von Eye (Ed.), *Statistical methods in longitudinal research* (Vol. 1). Boston: Academic Press.

GLASS, G. V., WILSON, V. L., AND GOTTMAN, J. M. (1975). *Design and analysis of time-series.* Colorado: Colorado University Press.

GOLDBERGER, A. S., AND DUNCAN, O. D. (1973). *Structural equations models in the social sciences.* New York: Seminar Press.

GOODMAN, L. A. (1984). *The analysis of cross-classified data having ordered catego-*

ries. Cambridge, MA: Harvard University Press.

GOODMAN, L. A., AND KRUSKAL, W. H. (1954). "Measures of association for cross-classifications," *Journal of the American Statistical Association.* **49**, 732–764.

HABERMAN, S. J. (1978). *Analysis of qualitative data (Vol. 1).* Introductory topics. New York: Academic Press.

HARTMANN, W. (1979). *Geometrische Modelle zur Analyse empirischer Daten.* Berlin (DDR): Akademie Verlag.

HAYDUK, L. (1987). *Structural equations modeling with LISREL.* Baltimore: Johns Hopkins University Press.

HERTZOG, C., AND NESSELROADE, J. R. (1987). "Beyond autoregressive models: Some implications of the trait-state distinction for structural equation modeling of developmental change," *Child Development,* **58**, 93–109.

HERTZOG, C., AND ROVINE, M. (1985). "Repeated-measures analysis of variance in developmental research: Selected issues," *Child Development,* **56**, 787–809.

HILDEBRAND, D. K., LAING, J. D., AND ROSENTHAL, H. (1977). *Prediction analysis of cross-classifications.* New York: Wiley.

HOAGLIN, D., MOSTELLER, F., AND TUKEY, J. (1983). *Understanding robust and exploratory data analysis.* New York: Wiley.

HOELTER, J. (1983). "The analysis of covariance structures: Goodness-of-fit indices," *Sociological Methods and Research,* **2**(3), 325–344.

HOLLAND, B. S., AND COPENHAVER, D. M. (1987). "An improved sequentially rejective Bonferroni test procedure," *Biometrics,* **43**, 417–423.

HOLM, S. (1979). "A simple sequentially rejective multiple test procedure," *Scandinavian Journal of Statistics,* **6**, 65–70.

JONES, C., AND NESSELROADE, J. (1989). *The multivariate, replicated, single subject repeated measures design and P-technique factor analysis: A review.* University Park, PA: The Pennsylvania State University.

JÖRESKOG, K. (1973). A general method for estimating a linear structural equation system. In A. S. Goldberger and O. D. Duncan (Eds.), *Structural equation models in the social sciences.* New York: Seminar Press.

JÖRESKOG, K., AND SÖRBOM, D. (1989). *LISREL VII: User's guide.* Los Angeles: SPSS Statistical Software.

JÖRESKOG, K., AND WOLD, H. (1982). *Systems under indirect observation: Causality, structure, prediction.* New York: North Holland.

KAPLAN, E. L., AND MEIER, P. (1958). "Nonparametric estimation from incomplete observations," *Journal of the American Statistical Association.* **53**, 457–481.

KESELMAN H. J. (1982). "Multiple comparisons for repeated measures means," *Multivariate Behavioral Research.* **17**, 87–92.

KIRK, R. E. (1982). *Experimental design.* Belmont: Brooks Cole.

KRAUSE, B., AND METZLER, P. (1984). *Angewandte Statistik.* Berlin (GDR): VEB Deutscher Verlag der Wissenschaften.

KRAUTH, J., AND LIENERT, G. A. (1973). *KFA. Die Konfigurations-frequenzanalyse und ihre Anwendung in Psychologie und Medizin.*

KRUSKAL, J. B. (1964a). "Multidimensional scaling by optimizing goodness of fit to a nonmetric hypothesis," *Psychometrika.* **29**, 1–27.

KRUSKAL, J. B. (1964b). "Nonmetric multidimensional scaling: A numerical method," *Psychometrika.* **29**, 28–42.

KRUSKAL, J. B., AND WISH, M. (1978). *Multidimensional scaling.* Beverly Hills: Sage.

KUBINGER, K. D. (1988). On a Rasch-model-based test for noncomputerized adaptive testing. In R. Langeheine and J. Rost (Eds.), *Latent trait and latent class models* (pp. 277–289). New York: Plenum.

KÜCHENHOFF, H. (1986). "A note on a continuity correction for testing in three-dimensional Configural Frequency Analysis," *Biometrical Journal.* **28**, 465–468.

LANGEHEINE, R. (1988). New developments in latent class theory. In R. Langeheine and J. Rost (Eds.), *Latent trait and latent class models,* (pp. 77–108). New York: Plenum.

LANGEHEINE, R., AND ROST, J. (Eds.) (1988). *Latent trait and latent class models.* New York: Plenum.

LARSEN, R. (1990). Spectral analysis. In A. von Eye (Ed.), *Statistical methods in longitudinal research* (Vol. 2). Boston: Academic Press.

LEHMACHER, W. (1981). "A more powerful simultaneous test procedure in Configural Frequency Analysis," *Biometrical Journal.* **23**, 429–436.

LIENERT, G. A., AND KRAUTH, J. (1973). "Die Konfigurationsfrequenzanalyse VI. Profiländerungen und Symptomverschiebungen," *Zeitschrift für Klinische Psychologie und Psychotherapie.* **21**, 100–109.

LITTLE R. J. A., AND RUBIN, D. B. (1987). *Statistical analysis with missing data.* New York: Wiley.

LINGOES, J. C. (1973). *The Guttman-Lingoes nonmetric program series.* Ann Arbor, MI: Mathesis Press.

LONG, J. S. (1983). *Covariance structure models: An introduction to LISREL.* Beverly Hills: Sage.

MACCALLUM, R. (1988). Multidimensional scaling. In J. R. Nesselroade and R. B. Cattell (Eds.), *Handbook of multivariate experimental psychology* (2nd ed.) (pp. 421–445). New York: Plenum.

MARINI, M. M., OLSEN A. R., AND RUBIN D. B. (1980). Maximum likelihood estimation in panel studies with missing data. *Sociological methodology.* San Francisco: Jossey Bass.

MAUCHLY, J. W. (1940). "Significance test for sphericity of a normal n-variate distribution," *Annals of Mathematical Statistics.* **11**, 204–209.

MCARDLE, J. J., AND ABER, M. S. (1990). Patterns of change within latent variable structural equation models. In A. von Eye (Ed.), *Statistical methods in longitudinal research* (Vol. 1). Boston: Academic Press.

MCARDLE, J. J., AND EPSTEIN, D. (1987) "Latent growth curves within developmental structural equation models," *Child Development.* **58**, 110–133.

MCARDLE, J. J., AND HORN, J. L. (1988). "An effective graphic model for linear structural models," *In Multivariate Behavioral Research* (in press).

MCARDLE, J. J., AND MCDONALD R. P. (1984). "Some algebraic properties of the Reticular Action Model for moment structures," *British Journal of Mathematical and Statistical Psychology.* **37**, 234–251.

MCARDLE, J. J., AND NESSELROADE, J. R. (1988). Structuring data to study development and change. In S. H. Cohen and H. W. Reese (Eds.), *Life-span developmental psychology: Methodological innovations.* Hillsdale, NJ: Erlbaum (in press).

MCCALL, R. B., AND APPELBAUM M. I. (1973). "Bias in the analysis of repeated-measures designs: Some alternative approaches," *Child Development.* **44**, 401–415.

MYERS, J. L. (1979). *Fundamentals of experimental design.* Boston: Allyn and Bacon.

METZLER, P., AND NICKEL, B. (1986). *Zeitreihen- und Verlaufsanalysen.* Leipzig: Hirzel.

MILLER, R. G. (1981). *Survival analysis.* New York: Wiley.

MITZEL, H., AND GAMES, P. (1981). "Circularity and multiple comparisons in repeated measure designs," *British Journal of Mathematical and Statistical Psychology.* **34**, 253–259.

MORRISON, D. F. (1976). *Multivariate statistical methods.* New York: McGraw-Hill.

NESSELROADE, J. R., AND BALTES, P. B. (1974). Adolescent personality development and historical change: 1970–1972 *Monographs of the Society for Research in Child Development.* **39**, No. 154.

NESSELROADE, J. R., AND BALTES, P. B. (1979). *Longitudinal research in the study of behavior and development.* New York: Academic Press.

NESSELROADE, J. R., PRUCHNO, R., AND JACOBS, A. (1986). "Reliability and stability in the measurement of psychological states: An illustration with anxiety measures," *Psychologische Beiträge.* **28**, 252–264.

NETER, J., WASSERMAN, W., AND KUTNER, W. (1985). *Applied linear statistical models.* Homewood, IL: Irwin.

PETERSEN, T. (1990). Analyzing event histories. In A. von Eye (Ed.), *Statistical methods in longitudinal research* (Vol. 1). Boston: Academic Press.

PUGESEK, B., AND WOOD P. (1989). *A longitudinal study of reproduction in the Cali-*

fornia Gull. The Pennsylvania State University, College of Health and Human Development, University Park, PA.

RAO, C. R. (1958). "Some statistical methods for the comparison of growth curves," *Biometrics.* **14**, 1–17.

RINDSKOPF, D. (1987). A compact BASIC program for loglinear models. In R. M. Heiberger (Ed.), *Computer science and statistics,* (pp. 381–386). Alexandria, VA: American Statistical Association.

RINDSKOPF, D. (1987). "Using latent class analysis to test developmental models," *Developmental Review.* **7**, 66–85.

RINDSKOPF, D. (1990). Testing developmental models using latent class analysis. In A. von Eye (Ed.), *Statistical methods in longitudinal research* (Vol. 2). Boston: Academic Press.

ROVINE, M. J., AND DELANEY, M. (1990). Missing data estimation in developmental research. In A. von Eye (Ed.), *Statistical methods in longitudinal research* (Vol. 1). Boston: Academic Press.

ROGOSA, D., BRANDT, D., AND ZIMKOWSKI, M. (1982). "A growth curve approach to the measurement of change," *Psychological Bulletin.* **90**, 726–748.

RUDINGER, G., CHASELON, F., ZIMMERMANN, E. J., AND HENNING, H. J. (1985). *Qualitative Daten. Neue Wege sozialwissenschaftli-cher Methodik.* München: Urban und Schwarzenberg.

SAS. (1985). Statistical Analysis System. Cary, NC: SAS Institute.

SCHIFFMAN, S. S., REYNOLDS, M. L., YOUNG, F. W. (1981). *Introduction to multidimensional scaling: Theory, methods, and applications.* New York: Academic Press.

SCHMITZ, B. (1990). Univariate and multivariate time series models: The analysis of intraindividual variability and intraindividual relationships. In A. von Eye (Ed.), *Statistical methods in longitudinal research* (Vol. 1). Boston: Academic Press.

SPSS, INC. (1983). *SPSSX user's guide.* New York: McGraw-Hill.

STEINBERG, D., AND COLLA, P. (1988). *SURVIVAL: A supplementary module for SYSTAT.* Evanston, IL: SYSTAT.

SURRA, C. (1988). *TUCKROT—A program for rotation of learning curves.* [computer program] Available from Catherine Surra, University of Arizona, School of Family and Consumer Resources, Tempe, AZ.

SZABAT, K. (1982). *Ex post Del.* Unpublished doctoral dissertation. University of Pennsylvania, Department of Statistics.

SZABAT, K. (1990). Prediction Analysis. In A. von Eye (Ed.), *Statistical methods in longitudinal research* (Vol. 2). Boston: Academic Press.

TAKANE, Y., YOUNG, F., AND DE LEEUW, J. (1976). "Nonmetric individual differences multidimensional scaling: An alternative least squares method with optimal scaling features," *Psychometrika.* **42**, 7–67.

THISSEN, D., AND BOCK, R. D. (1990). Linear and nonlinear curve fitting. In A. von Eye (Ed.), *Statistical methods in longitudinal research*. Boston: Academic Press.

THURSTONE L. L., AND THURSTONE T. G. (1962). *SRA primary mental abilities*. Chicago: Science Research Associates.

TISAK, J., AND MEREDITH, W. (1990). Longitudinal factor analysis. In A. von Eye (Ed.), *Statistical methods in longitudinal research* (Vol. 1). Boston: Academic Press.

TISAK, J., AND MEREDITH, W. (1990). Descriptive and associative developmental models. In A. von Eye (Ed.), *Statistical methods in longitudinal research* (Vol. 1). Boston: Academic Press.

TITTERINGTON, D. M., SMITH, A. F., AND MAKOV, U. E. (1985). *Statistical analysis of finite mixture distributions*. Chichester: Wiley.

TUCKER, L. (1966). Learning theory and multivariate experiment: Illustration of generalized learning curves. In R. B. Cattell (Ed.), *Handbook of multivariate experimental psychology*. New York: Rand McNally.

TUMA, N. B., AND HANNAN, M. T. (1984). *Social dynamics. Models and methods*. Orlando, FL: Academic Press.

TURNBULL, B. W. (1976). "The empirical distribution function with arbitrarily grouped, censored and truncated data," *Journal of the Royal Statistical Society*. **38**, 290–295.

UPTON, G. (1978). *The analysis of cross-tabulated data*. New York: Wiley.

VANDAELE, W. (1983). *Applied time series analysis and Box–Jenkins models*. Orlando: Academic Press.

VAN DRIEL, O. (1978). "On various causes of improper solutions in maximum likelihood factor analysis," *Psychometrika*. **43**(2), 225–243.

VON EYE, A. (1990a). Configural Frequency Analysis of longitudinal multivariate responses. In A. von Eye (Ed.), *Statistical methods in longitudinal research* (Vol. 2). Boston: Academic Press.

VON EYE, A. (1990b). *Introduction to Configural Frequency Analysis: The search for types and antitypes in cross-classifications*. Cambridge: Cambridge University Press.

VON EYE, A., AND BRANDTSTÄDTER, J. (1988a). Evaluating developmental hypotheses using statement calculus and nonparametric statistics. In P. B. Baltes, D. Featherman, and R. M. Lerner (Eds.), *Life-span development and behavior* (Vol. 8) (pp. 61–97). Hillsdale, NJ: Lawrence Erlbaum.

VON EYE, A., AND BRANDTSTÄDTER, J. (1988b). "Application of prediction analysis to cross-classifications of ordinal data," *Biometrical Journal*. **30**, 651–655.

VON EYE, A., AND KRAMPEN, G. (1987). "BASIC programs for prediction analysis of cross-classifications," *Educational and Psychological Measurement*. **47**, 141–143.

VON EYE, A., KREPPNER, K., AND WEβELS, H. (in prep.). Log-linear modeling of categorical data in developmental research.

VON EYE, A., AND NESSELROADE, J. R. (1990). "Types of change: Application of Configural Frequency Analysis to repeated observations in developmental research," *Experimental Aging Research.*

VON EYE, A., AND ROVINE, M. J. (1988). "A comparison of significance tests for Configural Frequency Analysis," *EDP in Medicine and Biology.* **19**, 6–13.

WILKINSON, L. (1988). *SYSTAT: The system for statistics.* Evanston, IL: SYSTAT, Inc.

WOHLWILL, J. (1973). *The study of behavioral development.* New York: Academic Press.

WOOD, P. K. (1990). Applications of scaling to developmental research. In A. von Eye (Ed.), *Statistical methods in longitudinal research* (Vol 2). Boston: Academic Press.

Author Index

Subject Index